T0203531

IET TELECOMMUNICATIONS SERIES 98

Artificial Intelligence Applied to Satellite-based Remote Sensing Data for Earth Observation

Other volumes in this series:

Volume 9	**Phase noise in signal sources** W.P. Robins
Volume 12	**Spread spectrum in communications** R. Skaug and J.F. Hjelmstad
Volume 13	**Advanced signal processing** D.J. Creasey (Editor)
Volume 19	**Telecommunications traffic, tariffs and costs** R.E. Farr
Volume 20	**An introduction to satellite communications** D.I. Dalgleish
Volume 26	**Common-channel signalling** R.J. Manterfield
Volume 28	**Very small aperture terminals (VSATs)** J.L. Everett (Editor)
Volume 29	**ATM: the broadband telecommunications solution** L.G. Cuthbert and J.C. Sapanel
Volume 31	**Data communications and networks, 3rd edition** R.L. Brewster (Editor)
Volume 32	**Analogue optical fibre communications** B. Wilson, Z. Ghassemlooy and I.Z. Darwazeh (Editors)
Volume 33	**Modern personal radio systems** R.C.V. Macario (Editor)
Volume 34	**Digital broadcasting** P. Dambacher
Volume 35	**Principles of performance engineering for telecommunication and information systems** M. Ghanbari, C.J. Hughes, M.C. Sinclair and J.P. Eade
Volume 36	**Telecommunication networks, 2nd edition** J.E. Flood (Editor)
Volume 37	**Optical communication receiver design** S.B. Alexander
Volume 38	**Satellite communication systems, 3rd edition** B.G. Evans (Editor)
Volume 40	**Spread spectrum in mobile communication** O. Berg, T. Berg, J.F. Hjelmstad, S. Haavik and R. Skaug
Volume 41	**World telecommunications economics** J.J. Wheatley
Volume 43	**Telecommunications signalling** R.J. Manterfield
Volume 44	**Digital signal filtering, analysis and restoration** J. Jan
Volume 45	**Radio spectrum management, 2nd edition** D.J. Withers
Volume 46	**Intelligent networks: principles and applications** J.R. Anderson
Volume 47	**Local access network technologies** P. France
Volume 48	**Telecommunications quality of service management** A.P. Oodan (Editor)
Volume 49	**Standard codecs: image compression to advanced video coding** M. Ghanbari
Volume 50	**Telecommunications regulation** J. Buckley
Volume 51	**Security for mobility** C. Mitchell (Editor)
Volume 52	**Understanding telecommunications networks** A. Valdar
Volume 53	**Video compression systems: from first principles to concatenated codecs** A. Bock
Volume 54	**Standard Codecs: image compression to advanced video coding, 3rd edition** M. Ghanbari
Volume 59	**Dynamic Ad Hoc Networks** H. Rashvand and H. Chao (Editors)
Volume 60	**Understanding Telecommunications Business** A Valdar and I Morfett
Volume 65	**Advances in Body-Centric Wireless Communication: Applications and State-of-the-art** Q. H. Abbasi, M. U. Rehman, K. Qaraqe and A. Alomainy (Editors)
Volume 67	**Managing the Internet of Things: Architectures, Theories and Applications** J. Huang and K. Hua (Editors)
Volume 68	**Advanced Relay Technologies in Next Generation Wireless Communications** I. Krikidis and G. Zheng
Volume 69	**5G Wireless Technologies** A. Alexiou (Editor)
Volume 70	**Cloud and Fog Computing in 5G Mobile Networks** E. Markakis, G. Mastorakis, C. X. Mavromoustakis and E. Pallis (Editors)
Volume 71	**Understanding Telecommunications Networks, 2nd edition** A. Valdar
Volume 72	**Introduction to Digital Wireless Communications** Hong-Chuan Yang
Volume 73	**Network as a service for next generation internet** Q. Duan and S. Wang (Editors)
Volume 74	**Access, Fronthaul and Backhaul Networks for 5G & Beyond** M. A. Imran, S. A. R. Zaidi and M. Z. Shakir (Editors)
Volume 76	**Trusted Communications with Physical Layer Security for 5G and Beyond** T. Q. Duong, X. Zhou and H. V. Poor (Editors)
Volume 77	**Network Design, Modelling and Performance Evaluation** Q. Vien
Volume 78	**Principles and Applications of Free Space Optical Communications** A. K. Majumdar, Z. Ghassemlooy, A. A. B. Raj (Editors)
Volume 79	**Satellite Communications in the 5G Era** S. K. Sharma, S. Chatzinotas and D. Arapoglou
Volume 80	**Transceiver and System Design for Digital Communications 5th Edition** Scott R. Bullock
Volume 81	**Applications of Machine Learning in Wireless Communications** R. He and Z. Ding (Editors)
Volume 83	**Microstrip and Printed Antenna Design 3rd Edition** R. Bancroft
Volume 84	**Low Electromagnetic Emission Wireless Network Technologies: 5G and beyond** M.A. Imran, F. Héliot and Y.A. Sambo (Editors)
Volume 86	**Advances in Communications Satellite Systems Proceedings of The 36th International Communications Satellite Systems Conference (ICSSC-2018)** I. Otung, T. Butash and P. Garland (Editors)
Volume 89	**Information and Communication Technologies for Humanitarian Services** M. N. Islam (Editor)
Volume 90	**Communication Technologies for Networked Smart Cities** S.K. Sharma, N. Jayakody, S. Chatzinotas and A. Anpalagan (Editors)
Volume 91	**Green Communications for Energy-Efficient Wireless Systems and Networks** Himal Asanga Suraweera, Jing Yang, Alessio Zappone and John S. Thompson (Editors)

Volume 92   **Flexible and Cognitive Radio Access Technologies for 5G and Beyond** H. Arslan and E. Başar (Editors)
Volume 93   **Antennas and Propagation for 5G and Beyond** Q. Abbasi, S.F. Jilani, A. Alomainy and M. A. Imran (Editors)
Volume 94   **Intelligent Wireless Communications** G. Mastorakis, C. X. Mavromoustakis, J. M. Batalla and E. Pallis (Editors)
Volume 95   **ISDN applications in education and training** R. Mason and P.D. Bacsich

Artificial Intelligence Applied to Satellite-based Remote Sensing Data for Earth Observation

Edited by
Maria Pia Del Rosso, Alessandro Sebastianelli
and Silvia Liberata Ullo

The Institution of Engineering and Technology

Published by The Institution of Engineering and Technology, London, United Kingdom

The Institution of Engineering and Technology is registered as a Charity in England & Wales (no. 211014) and Scotland (no. SC038698).

© The Institution of Engineering and Technology 2021

First published 2021

The Institution of Engineering and Technology
Michael Faraday House
Six Hills Way, Stevenage
Herts, SG1 2AY, United Kingdom

www.theiet.org

British Library Cataloguing in Publication Data
A catalogue record for this product is available from the British Library

ISBN 978-1-83953-212-2 (hardback)
ISBN 978-1-83953-213-9 (PDF)

Typeset in India by Exeter Premedia Services Private Limited
Printed in the UK by CPI Group (UK) Ltd, Croydon

Contents

Mathematical notation   xv
List of figures   xvii
List of tables   xxiii
About the editors   xxv

1   **The rise of Artificial Intelligence (AI) for Earth Observation (EO)**   1
     Pierre Philippe Mathieu

     1.1   Embracing external trends   1
     1.2   The Artificial Intelligence and Deep Learning revolution   2
     1.3   Rise of Artificial Intelligence for Earth Observation   2
         1.3.1   AI4EO data engine   3
         1.3.2   AI@edge   4
     1.4   Conclusions   7

2   **Principles of satellite data analysis**   9
     Alessandro Sebastianelli, Silvia Liberata Ullo, Maria Pia Del Rosso,
     Sara Aparício, Andrea Radius, Chiara Zarro, James Wheeler,
     Chris Stewart, and Pia Addabbo

     2.1   Introduction   9
         2.1.1   Principles underlying remote sensing   10
             2.1.1.1   Electromagnetic radiation   10
             2.1.1.2   Electromagnetic spectrum   11
         2.1.2   Interaction between electromagnetic waves and matter   12
             2.1.2.1   Reflection   12
             2.1.2.2   Refraction   13
             2.1.2.3   Diffraction   14
         2.1.3   Other types of interaction   14
         2.1.4   Interaction with the atmosphere   16
             2.1.4.1   Scattering   17
             2.1.4.2   Absorption   18
             2.1.4.3   Refraction   18
     2.2   Types of sensors   18
         2.2.1   Active sensors   18
         2.2.2   Passive sensors   20
     2.3   Platforms   21
         2.3.1   Ground-based platforms   21

	2.3.2	Airborne platforms	22
	2.3.3	Spaceborne platforms	23
2.4	Data characteristics		23
	2.4.1	Operations for improving quality of data	25
2.5	Copernicus programme		26
	2.5.1	The Sentinels	26
2.6	Focus on optical and SAR data		27
	2.6.1	Principles of SAR	28
	2.6.2	Classic processing workflows for SAR data	29
	2.6.3	Principles of optical sensors	31
	2.6.4	Classic processing workflows for optical data	31
	2.6.5	Applications	32
		2.6.5.1 An application of SAR data: the interferometry	32
		2.6.5.2 An application of optical data: the object-based image analysis	33
2.7	Conclusions		35
	References		35

3   Artificial intelligence, machine learning and deep learning   **39**
Maria Pia Del Rosso, Silvia Liberata Ullo, Alessandro Sebastianelli,
Dario Spiller, Erika Puglisi, Diego Di Martire, Sara Aparicio, and
Pia Addabbo

3.1	Introduction		39
	3.1.1	Historical pills on the birth of AI	40
3.2	Artificial intelligence, machine learning and deep learning		43
	3.2.1	Machine learning	43
	3.2.2	Deep learning	43
	3.2.3	Real-life applications	44
3.3	Supervised, unsupervised and reinforcement learning		45
	3.3.1	Supervised learning	45
	3.3.2	Unsupervised learning	47
	3.3.3	Reinforcement learning	49
3.4	Training and validation of a model		50
3.5	Cross-validation for supervised learning		50
3.6	Class imbalance and data augmentation		51
	3.6.1	Class imbalance	52
	3.6.2	Data augmentation	53
		3.6.2.1 For tabular data	53
		3.6.2.2 For image data	54
	3.6.3	Noise robustness	54
	3.6.4	Choice and use of a dataset	54
3.7	Applications		55
	3.7.1	Remote sensing applications	55
		3.7.1.1 Natural disasters	55

    3.7.2   Other applications 57
          3.7.2.1   Self-driving cars 57
          3.7.2.2   E-commerce and social media 57
          3.7.2.3   Space exploration 57
          3.7.2.4   Fault detection of navigation systems 57
          3.7.2.5   Medical applications 58
3.8   Conclusions 58
References 58

4 Artificial neural network 63
Maria Pia Del Rosso, Silvia Liberata Ullo, Alessandro Sebastianelli,
Dario Spiller, Erika Puglisi, Filippo Biondi, and Danilo Orlando

4.1   Introduction 63
4.2   Artificial neural networks 63
    4.2.1   Artificial neuron 65
4.3   Activation functions 65
    4.3.1   Sigmoid function 67
    4.3.2   Hyperbolic tangent function (tanh) 67
    4.3.3   Rectified linear unit function 68
    4.3.4   The bias 69
4.4   Forward and backward propagation 72
    4.4.1   The gradient descent 72
    4.4.2   Description 73
    4.4.3   Stochastic gradient descent 75
    4.4.4   Full batch gradient descent 76
    4.4.5   Mini-batch gradient descent 76
          4.4.5.1   Tips 77
    4.4.6   Batch normalisation 77
    4.4.7   The vanishing gradient 78
          4.4.7.1   Solutions 78
    4.4.8   The gradient explosion 79
          4.4.8.1   Solution 79
    4.4.9   The back propagation 79
    4.4.10 A numerical example 80
4.5   Overfitting and underfitting 84
    4.5.1   Bias and variance 84
    4.5.2   Early stopping of training 85
    4.5.3   Dropout 86
4.6   Regularisation techniques 87
    4.6.1   Parameter norm penalties 87
          4.6.1.1   Regularisation L2 87
          4.6.1.2   Regularisation L1 88
4.7   Conclusions 88
References 88

5   Convolutional neural networks   91
Silvia Liberata Ullo, Alessandro Sebastianelli, Maria Pia Del Rosso,
Dario Spiller, Erika Puglisi, Artur Nowakowski, Mario Luca Bernardi,
and Marta Cimitile

   5.1   Introduction   91
   5.2   Convolutional neural networks   93
       5.2.1   Motivation   93
       5.2.2   The convolution operation   95
       5.2.3   Parameter sharing and sparse connectivity   97
   5.3   Networks and layers structure   99
   5.4   Kernels and feature maps   100
       5.4.1   Stride   102
       5.4.2   Multidimensional input   103
       5.4.3   Padding   103
   5.5   ReLU layer   104
   5.6   Pooling   106
   5.7   Batch normalisation   107
       5.7.1   Motivation   107
       5.7.2   How does batch normalisation work?   108
   5.8   Common tasks performed by CNNs   108
   5.9   Conclusions   109
   References   109

6   How to create a proper EO dataset   113
Maria Pia Del Rosso, Alessandro Sebastianelli, Sara Aparicio, and
James Wheeler

   6.1   Introduction   113
   6.2   Non-automated method   114
       6.2.1   The open access hub   114
       6.2.2   SNAP – Sentinel Application Platform   116
   6.3   Automated method   117
       6.3.1   Google Earth Engine – online editor   117
       6.3.2   Google Earth Engine – Python API   119
   6.4   Dataset annotation   120
   6.5   Training dataset for image classification tasks   121
       6.5.1   Annotation tool for image classification – makesense.ai   121
   6.6   Training dataset for semantic segmentation tasks   123
       6.6.1   Annotation tool for semantic segmentation – Google Earth
            Engine online editor   125
   6.7   Training dataset for object detection tasks   127
       6.7.1   Annotation tool for semantic segmentation – LabelImg   128

6.8   Conclusions   130
References   130

7   How to develop your network with Python and Keras   131
Silvia Liberata Ullo, Maria Pia Del Rosso, Alessandro Sebastianelli,
Erika Puglisi, Mario Luca Bernardi, and Marta Cimitile

7.1   Introduction   131
      7.1.1   TensorFlow and Keras   131
      7.1.2   Tensors   132
7.2   Google Colaboratory   134
      7.2.1   How to work with GPU?   136
      7.2.2   How to connect Google Drive?   137
7.3   Examples   138
      7.3.1   Noise filtering   144
              7.3.1.1   Step 1 – Define a generator for the dataset   145
              7.3.1.2   Step 2 – Build a model   145
              7.3.1.3   Step 3 – Train and test the model   147
      7.3.2   Classification   148
              7.3.2.1   Step 1 – Define a generator for the dataset   148
              7.3.2.2   Step 2 – Build a model   150
              7.3.2.3   Step 3 – Train and test the model   151
      7.3.3   Detection   153
              7.3.3.1   Step 1 – Define a generator for the dataset   153
              7.3.3.2   Step 2 – Build a model   154
              7.3.3.3   Step 3 – Train and test the model   155
7.4   Conclusions   157
References   157

8   A classification problem   159
Alessandro Sebastianelli, Maria Pia Del Rosso, Chiara Zarro,
Diego Di Martire, Mariano Di Napoli, Dario Spiller, Artur Nowakowski,
Rogerio Bonifacio, Do-Hyung Kim, and Manuel Garcia-Herranz

8.1   Introduction   159
8.2   Case study A: volcanic eruptions binary classification   159
      8.2.1   The catalog   159
      8.2.2   Data preparation and manipulation   160
      8.2.3   Dataset expansion   161
      8.2.4   Proposed model   162
              8.2.4.1   Image Loader   165
      8.2.5   Training   165
      8.2.6   Test on test dataset   167
      8.2.7   Future works   169

8.3	Case Study B: Landslides binary classification	171
	8.3.1 Landslide classification	173
	8.3.2 The CNN for landslides binary classification	176
	8.3.3 The method based on optical images	176
	8.3.4 SAR data and creation of the dataset	180
	8.3.4.1 GRD and SLC SAR data	181
	8.3.5 The proposed CNN for SAR data	184
	8.3.6 Results	186
	8.3.7 Future developments	189
	8.3.7.1 Dataset maintenance	190
	8.3.7.2 RGB composite with SLC data	190
8.4	Case Study C: Crops-type classification	191
	8.4.1 CTC problem	191
	8.4.2 Dataset	192
	8.4.3 Classification using CNNs and transfer learning	193
	8.4.4 Results	194
	8.4.5 Future works	195
8.5	Case Study D: Segmentation	195
	8.5.1 Segmentation tool and algorithms	196
	8.5.2 Experimental analysis and results	196
	8.5.2.1 Contrast split segmentation	196
	8.5.2.2 Multiresolution segmentation	197
	8.5.3 Integration of deep learning and object-based analysis	199
8.6	Conclusions	199
	References	200

9   A generation problem   **207**
Alessandro Sebastianelli, Maria Pia Del Rosso, Silvia Liberata Ullo,
Erika Puglisi, and Filippo Biondi

9.1	Introduction	207
9.2	Deep generative models	208
	9.2.1 Variational autoencoders	208
	9.2.2 Generative adversarial networks	210
9.3	Case study: Generation of optical-like images	211
	9.3.1 Dataset and proposed methods	212
9.4	Model for the generation of optical-like data	213
	9.4.1 Generator	213
	9.4.2 Discriminator	213
	9.4.3 Loss function selection	213
9.5	Results	214
	9.5.1 Test on simulated Sentinel-1 images	215
	9.5.2 Test on real Sentinel-1 images	215
9.6	Conclusions	215
	References	219

10 A filtering problem: SAR speckle filtering **221**
Alessandro Sebastianelli, Maria Pia Del Rosso, Silvia Liberata Ullo,
Andrea Radius, Carmine Clemente, Danilo Orlando, Filippo Biondi,
and James Wheeler

10.1 Introduction 221
    10.1.1 Speckle filtering motivation 222
    10.1.2 Speckle reduction 223
    10.1.3 Speckle filtering examples 224
10.2 Case study: Sentinel-1 speckle filtering 225
    10.2.1 Multi-modal speckle filtering 225
    10.2.2 Pure SAR strategy 226
    10.2.3 CNN model for SAR speckle filtering 226
    10.2.4 Network architecture 227
    10.2.5 Loss function selection 228
10.3 Evaluation metrics 228
    10.3.1 Peak signal-to-noise ratio 228
    10.3.2 The structural similarity index 229
10.4 Results 229
    10.4.1 Data preparation 230
    10.4.2 Test on small patches of a Sentinel-1 image 231
        10.4.2.1 Test on the training dataset 231
        10.4.2.2 Test on validation dataset 231
        10.4.2.3 Comparison with the state of the art 232
    10.4.3 Test on a whole Sentinel-1 image 233
10.5 Conclusions 234
References 235

11 Future perspectives and conclusions **237**
Silvia Liberata Ullo

11.1 Introduction 237
11.2 A first prototype 238
    11.2.1 Raspberry Pi 238
    11.2.2 Movidius Stick 239
    11.2.3 Camera 240
    11.2.4 Proposed model 240
    11.2.5 Results 241
    11.2.6 Implementation on Raspberry and Movidius Stick 241
    11.2.7 OpenVINO library 243
11.3 Future satellite missions with AI on board 244
    11.3.1 ESA first intelligent satellites: ϕ-Sat 1 and 2 245
11.4 Conclusions 246
11.5 Acknowledgements 246
References 247

Index **249**

Mathematical notation

Mathematical preliminary concepts

It is useful to state some rules of mathematical notation that will be used in this book, for describing the basic concepts of artificial neural networks.

Mathematical symbols are shown with italic fonts, and vectors are emphasised using bold fonts. So, for instance, x is a scalar quantity and $\mathbf{x}$ is a vector. Scalar variables can live in the real set $\mathbb{R}$ or in the natural set $\mathbb{N}$, whereas vectors are usually defined in $\mathbb{R}^n$, where n is the dimension of the vector space. By definition, $\mathbf{x} \in \mathbb{R}^n$ is given as a *column* vector, and its transpose $\mathbf{x}^T$ is a row vector.

The norm of a vector is a function defined over a vector space to map a vector into a scalar quantity. Given the generic p-norm definition,

$$\| \mathbf{x} \|_p = \left(x_1^p + x_2^p + \ldots + x_n^p \right)^{\frac{1}{p}}, \tag{1}$$

In this book the 1-norm:

$$\| \mathbf{x} \|_1 = | x_1 | + | x_2 | + \ldots + | x_n | \tag{2}$$

and the 2-norm:

$$\| \mathbf{x} \|_2 = \left(x_1^2 + x_2^2 + \ldots + x_n^2 \right)^{\frac{1}{2}} \tag{3}$$

will be used. The 2-norm is also known as the Euclidean norm and evaluates the intuitive concept of the length for two-dimensional and three-dimensional vectors.

A vector with unit 2-norm is referred to using the *hat* symbol, that is $\hat{\mathbf{x}}$ is defined as having $\| \hat{\mathbf{x}} \|_2 = 1$.

The inner product or scalar product is defined as an operation mapping two vectors in $\mathbb{R}^n$, for example $\mathbf{x}$ and $\mathbf{y}$, into $a \in \mathbb{R}$. It is defined as:

$$\mathbf{x}^T \mathbf{y} = x_1 y_1 + x_2 y_2 + \ldots + x_n y_n = a. \tag{4}$$

A two-dimensional matrix is a *table* of numbers with m rows and n columns. It is usually denoted by bold capital letters, not-italic symbols, for example $\mathbf{A}$. A two-dimensional $M \times N$ matrix is composed of elements $a\_\{ij\}$, $i \in [1, M]$, $j \in [1, N]$. The product of matrices is usually referred to as row-by-column multiplication. Given a $M_1 \times N$ matrix $\mathbf{A}$ and an $N \times M_2$ matrix $\mathbf{B}$, the product matrix $\mathbf{C}$ is given as

$$c_{ij} = a_{i1} b_{1j} + a_{i2} b_{2j} + \cdots + a_{in} b_{nj} = \sum_{k=1}^{n} a_{ik} b_{kj}. \tag{5}$$

A two-dimensional matrix can be used to map a vector $\mathbf{x}$ in $\mathbb{R}^n$ into a vector $\mathbf{y}$ in $\mathbb{R}^m$, i.e.

$$\mathbf{y} = \mathbf{A}\mathbf{x}. \tag{6}$$

List of figures

Figure 1.1   Illustration of an adaptive data engine for EO data based on Machine Learning. It is derived from a new type of value chain, whereby the algorithms are reverse engineered by learning from the data as opposed to the traditional way of explicitly coding knowledge. In AI-powered value chains, most of the work is made up-front (e.g. data preparation, learning) and the result comes from the application of a trained ML scheme (e.g. inference). The process of transforming raw data into information becomes thereby automatic, optimised and scalable, keeping in mind that the quality of the output is related to the quality of the input   4

Figure 1.2   Illustration of the 'pruning' of neural network   5

Figure 1.3   Digital Twin Earth framework   6

Figure 2.1   Electromagnetic wave   10

Figure 2.2   Electromagnetic spectrum   12

Figure 2.3   Reflection   13

Figure 2.4   Refraction   14

Figure 2.5   Diffraction   15

Figure 2.6   Doppler effect   15

Figure 2.7   Different types of polarisation   16

Figure 2.8   Example of active sensor   19

Figure 2.9   Example of passive sensor   19

Figure 2.10   Remote sensing platform   22

Figure 2.11   Spatial resolution   24

Figure 2.12   Spectral resolution   24

Figure 2.13   Radiometric resolution   25

Figure 3.1   Participants of the 1956 Dartmouth Summer Research Project on AI in front of Dartmouth Hall. In the photo there are also Claude Shannon (front right), John McCarthy (back right), Marvin Minsky (back centre) and Nathaniel Rochester (back left). Photo credit Margaret Minsky via www.achievement.org.   41

Figure 3.2   Difference between a simple neural network and a deep neural network   42

Figure 3.3   Differences and relationships between AI, ML and DL   43

Figure 3.4   Some of the possible AI applications   44

Figure 3.5   Typical examples of algorithms for supervised learning   46

Figure 3.6   Typical examples of unsupervised learning algorithms   48

Figure 3.7	Definition of training, validation and test datasets	51
Figure 3.8	Example of ten-fold cross-validation	52
Figure 4.1	An artificial neural is an interconnected group of nodes, inspired by the simplification of neurons in the brain. Each circle represents an artificial neuron and an arrow represents a connection (i.e. an edge) from the output of one artificial neuron to the input of another	64
Figure 4.2	Example of an artificial neuron	65
Figure 4.3	Step and linear activation functions	66
Figure 4.4	Sigmoid function	67
Figure 4.5	Tanh function	68
Figure 4.6	The rectified linear unit function (ReLU) function	69
Figure 4.7	The leaky rectified linear unit function (ReLU) function	69
Figure 4.8	A one-input one-output neural network	70
Figure 4.9	Graphic representation of results of a one-input one-output neural network without bias with different values of the weight w, using the sigmoid as the activation function	70
Figure 4.10	A one-input one-output neural network with a bias neuron	71
Figure 4.11	Graphic representation of results of a one-input one-output neural network with different values of bias and fixed weight w, using the sigmoid as the activation function	71
Figure 4.12	An illustration of how the gradient descent algorithm uses the derivatives of a function to reach a minimum	73
Figure 4.13	Optimisation algorithms may not find a global minimum when the function returns multiple local minimum points. In the context of deep learning, we generally accept such solutions even if they do not belong to the absolute minima. This approximation is valid as long as solutions correspond to significantly low values of the cost function (these values should be comparable with the absolute minima)	74
Figure 4.14	Neural network as box	80
Figure 4.15	Neural networks step by step	80
Figure 4.16	Early stopping example	86
Figure 5.1	An example of convolutional neural networks (CNNs) that classify Sentinel-2 images	92
Figure 5.2	Sparse connectivity, where the input x_2 is highlighted as well as the output units s that are affected by this unit. (Left) When the s layer is formed by convolution with a kernel of width 3, only three outputs are affected by x. (Right) When the s layer is formed by matrix multiplication, connectivity is no longer sparse, so all of the outputs are affected by x_2	94
Figure 5.3	The receptive field of the units in the deeper layers of a convolutional network is larger than the receptive field of the units in the shallow layers. This effect increases if the network includes architectural features like strided convolution	

	or pooling. This means that even though direct connections in a convolutional net are very sparse, units in the deeper layers can be indirectly connected to all or most of the input image	94
Figure 5.4	Example of convolution between two squared pulses	96
Figure 5.5	Parameter sharing: Red arrows indicate the connections that use a particular parameter in two different models. (Left) The red arrows indicate uses of the central element of a three-element kernel in a convolutional model. Due to parameter sharing, this single parameter is used at all input locations. (Right) The red arrow indicates the use of the second element of the weight matrix in a fully connected model. This model has no parameter sharing so the parameter is used only once	97
Figure 5.6	Efficiency of edge detection	98
Figure 5.7	The components of a typical convolutional neural network layer	99
Figure 5.8	Example of how the convolution operation is performed on an image. The matrix multiplication between the filter (the 5×5×3 square on the left) and a part of the input image forms an element (the circle on the right) of the feature map (represented in dark blue on the right). In this image ten different filters are used, this is why the output image has a size of $32 \times 32 \times 10$	100
Figure 5.9	First three filters of the first layer of the VGG16 network [13], one row for each filter and one column for each channel	101
Figure 5.10	The original image (in RGB on the left) and the feature maps generated by the filters of the VGG16 network (on the right). In the first row it is possible to see three feature maps generated by the first layer of the network (in particular the three ones generated by the three filters represented in Figure 5.9); in the second row the feature maps generated by the ninth layer and in the last row the feature maps generated by the sixteenth layer	102
Figure 5.11	Movement of the kernel represented by the blue line	103
Figure 5.12	Convolution operation with stride length = 1	103
Figure 5.13	Convolution operation on an $M \times N \times 3$ image matrix with a $3 \times 3 \times 3$ kernel	104
Figure 5.14	Same padding: $6 \times 6 \times n$ image is padded with a filter to create a $4 \times 4 \times n$ image	105
Figure 5.15	A visual presentation of how different kinds of pooling downsample the pixel values	107
Figure 6.1	The ESA open access hub – home page (source [5])	115
Figure 6.2	The ESA open access hub – select area of interest (AOI) (source [5])	115
Figure 6.3	The ESA open access hub – select settings (source [5])	116
Figure 6.4	The ESA open access hub – download data (source[1])	116
Figure 6.5	Sentinel Application Platform	117
Figure 6.6	Google Earth Engine code editor	118

Figure 6.7   Google Earth Engine code editor – Sentinel-2 data   118
Figure 6.8   Example of training dataset for image classification: labels
              (in vector/tabular format) are assigned to each image   122
Figure 6.9   makesense.ai: online labelling software   122
Figure 6.10   makesense.ai: defining the labels   123
Figure 6.11   makesense.ai: defining the labels   123
Figure 6.12   makesense.ai: exporting the labels   124
Figure 6.13   An example of a training dataset for an application aimed
              at classifying per pixel the land cover, collecting pixels from
              the image   124
Figure 6.14   An example of a training dataset for an application aimed
              at classifying per pixel, the land cover creating a mask from
              the image   125
Figure 6.15   Creation of classes to be sampled   126
Figure 6.16   Editing class properties   126
Figure 6.17   Picking pixels of different classes   127
Figure 6.18   Example of a training dataset for objection detection aimed
              at detecting water bodies   128
Figure 6.19   Image labelling software   129
Figure 6.20   Creating bounding boxes with class 'plane' on Sentinel-2 data   129
Figure 7.1   Graphic representation of a three-axes tensor of shape (3,2,5)   134
Figure 7.2   Google Colab GUI (Source [5])   135
Figure 7.3   Connect the GPU (Step 1) (Source [5])   136
Figure 7.4   Connect the GPU (Step 2) (Source [5])   137
Figure 7.5   Connect Google Drive (Step 1) (Source [4])   138
Figure 7.6   Connect Google Drive (Step 2) (Source [4])   138
Figure 7.7   Example of input and output produced by the generator   146
Figure 7.8   Example of model output   148
Figure 7.9   Example of input and output produced by the generator   150
Figure 7.10   Example of model output   152
Figure 7.11   Confusion matrix   153
Figure 7.12   Example of input and output produced by the generator   155
Figure 7.13   Example of model output   156
Figure 8.1   True color image vs. infrared highlights image   161
Figure 8.2   Smoke vs. cloud   162
Figure 8.3   A set of nine images from the dataset   163
Figure 8.4   Network architecture   164
Figure 8.5   Accuracy and loss functions   166
Figure 8.6   Confusion matrix   167
Figure 8.7   Test on test dataset (1)   168
Figure 8.8   Test on test dataset (2)   169
Figure 8.9   Test on test dataset (3)   170
Figure 8.10   Example of information acquired by Sentinel-5P during a
              volcanic eruption. Cit: https://earth.esa.int/web/sentinel/news/
              content/-/asset_publisher/BZewkR1itkH2/content/copernicus-

sentinel-5p-spots-so2-emissions title: Copernicus Sentinel-5P spots SO$_2$ emissions. Description: On May 6, 2018, Sentinel-5P and VIIRS (instrument on the NASA/NOAA satellite Suomi-NPP) detected sulfur dioxide emissions and lava (hot spots) from the Kīlauea volcano in Hawaii. Copyright: BIRA/IASB   172

Figure 8.11   Main landslides types with related Sentinel-1 and Sentinel-1 images (from top to down): debris flow at Xinmo (China), on June 24, 2017; debris slide at Piz Cengalo (Switzerland), on August 23, 2017; Rock fall at El Capitan in the Yosemite National Park, California (United States), on September 28, 2017; Earth flow at Morton Road, Dutch Flat, California (United States), on January 11, 2017   175

Figure 8.12   CNN s, pixel-based algorithms   176
Figure 8.13   Block scheme of the proposed CNN for Sentinel-2 images   178
Figure 8.14   Sentinel-2 and Landsat bands [37]   179
Figure 8.15   An example of Sentinel-2 images acquired before (a) and after (b) a landslide (Xinmo landslide, China, June 24, 2017)   179
Figure 8.16   Training and evaluation accuracy   180
Figure 8.17   Plot of the landslides under analysis (NASA Open Data GLC [26])   183
Figure 8.18   Sentinel-1 SAR image, acquired in VV polarization, of the same landslide of Figure 8.15 (zoom) (Xinmo landslide, China, June 24, 2017)   183
Figure 8.19   Proposed CNN architecture for SAR data   184
Figure 8.20   Training and validation accuracy (top). Training and validation loss (bottom).   187
Figure 8.21   Confusion matrix network 1   187
Figure 8.22   Confusion matrix network 2   188
Figure 8.23   Confusion matrix network 3   188
Figure 8.24   Change detection for the dataset creation   190
Figure 8.25   RGB composite summary scheme   191
Figure 8.26   Localizations of the orthomosaics (a–b) and selected orthomosaic with perimeters of reference fields (c)   192
Figure 8.27   Architectures of CNNs used in the study   193
Figure 8.28   Classification results using TL   194
Figure 8.29   Contrast split segmentation on NDSM layer   197
Figure 8.30   Multiresolution segmentation on WV-2 images layer   198
Figure 9.1   Example of an autoencoder structure   208
Figure 9.2   Example of how the dataset is distributed in the space from a classical autoencoder (on the left) and from a VAE (on the right). In the first case the figure in purple taken as input will generate an unrealistic output.   209
Figure 9.3   Example of a GAN architecture   211
Figure 9.4   Samples from the dataset   212
Figure 9.5   Sentinel-2 data generation   212

Figure 9.6	Network architecture for the generator	214
Figure 9.7	Network architecture for the discriminator	214
Figure 9.8	Test of the model on simulated Sentinel-1 images (1)	216
Figure 9.9	Test of the model on simulated Sentinel-1 images (2)	216
Figure 9.10	Test of the model on simulated Sentinel-1 images (3)	217
Figure 9.11	Test of the model on real Sentinel-1 images (1)	217
Figure 9.12	Test of the model on real Sentinel-1 images (2)	218
Figure 9.13	Test of the model on real Sentinel-1 images (3)	218
Figure 10.1	Example of a synthetic aperture radar (SAR) image with speckle (Rome, VV polarisation)	222
Figure 10.2	Comparison of speckle filters (source [6])	224
Figure 10.3	Generation of Sentinel-1 like images	226
Figure 10.4	Generation of Sentinel-1 images with speckle	227
Figure 10.5	Network architecture for synthetic aperture radar (SAR) image despeckling	227
Figure 10.6	Model loss	230
Figure 10.7	Example of the dataset	231
Figure 10.8	Test on training dataset	232
Figure 10.9	Test on validation dataset	233
Figure 10.10	256×256 batch of a Sentinel-1 product	234
Figure 10.11	Original Sentinel-1 product with real speckle	235
Figure 10.12	Filtered Sentinel-1 product without speckle	235
Figure 11.1	Raspberry Pi 3 B [1]	238
Figure 11.2	Movidius stick [2]	239
Figure 11.3	Raspberry RGB camera [3]	240
Figure 11.4	Accuracy and loss functions for the small model	242
Figure 11.5	Confusion matrix for the small model	242
Figure 11.6	Block diagram for the implementation on chip	243
Figure 11.7	φ-Sat chip [7]	245

List of tables

Table 4.1	Back propagation: forward propagate after initialisation	81
Table 4.2	Back propagation: forward propagation after initialisation	81
Table 4.3	Back propagation: loss functions	82
Table 4.4	Back propagation: loss functions	83
Table 7.1	Google Colab GPU specifications	137
Table 8.1	Sample from the catalog	160
Table 8.2	Image correction parameters	161
Table 8.3	Results on test dataset (1)	167
Table 8.4	Results on test dataset (2)	168
Table 8.5	Results on test dataset (3)	170
Table 8.6	A sample of landslides of interest	182
Table 8.7	Results of each test sorted by relevance	189
Table 8.8	Statistics on reference fields	192
Table 10.1	Comparison of ENL, Robert's Edge Operator and Variance Texture Operator of the original image and the images filtered using the Lee, Frost, Lee-Sigma, Gamma-Map. A 5×5 window was used in all cases (source [6])	225
Table 10.2	Evaluation metrics on training datasets	232
Table 10.3	Evaluation metrics on validation datasets	233
Table 10.4	Evaluation metrics on validation datasets	234
Table 11.1	Raspberry PI 3 specifications	239
Table 11.2	Raspberry Pi RGB camera features	241

About the editors

Maria Pia Del Rosso is an electronics engineer and researcher in the Engineering Department of the University of Sannio, Italy, with a particular interest in Remote Sensing and Artificial Intelligence. While a master student she worked on monitoring natural disaster through satellite data. She is a member of the IEEE Geoscience and Remote Sensing Society and the IEEE Communications Society.
ORCID: https://orcid.org/0000-0002-0297-0102

Alessandro Sebastianelli is an electronics engineer and researcher in the Engineering Department of the University of Sannio, Italy, with research interests in Remote Sensing and Artificial Intelligence. During his years as a master student he worked on monitoring natural disaster through satellite data. He is a member of the IEEE Communications Society and the IEEE Geoscience and Remote Sensing Society.
ORCID: https://orcid.org/0000-0002-9252-907X

Silvia Liberata Ullo is a professor and researcher in the Telecommunications and Remote Sensing group at the University of Sannio, Italy. Her research deals with data analysis and remote sensing techniques. She is a senior member of the IEEE and is associated with IEEE Aerospace and Electronic Systems, IEEE Geoscience and Remote Sensing, IEEE Communication, IEEE Signal Processing and IEEE Women In Engineering (WIE).
ORCID: https://orcid.org/0000-0001-6294-0581

Chapter 1

The rise of Artificial Intelligence (AI) for Earth Observation (EO)

Pierre Philippe Mathieu[1]

This chapter introduces the topic on AI4EO (Artificial Intelligence for Earth Observation).

1.1 Embracing external trends

Powerful trends in digital and sensing technologies are rapidly increasing, thereby also transforming the world of EO[a].

In particular, extraordinary developments in information and communication technologies (ICT), including the Internet, cloud computing and AI, are giving rise to radically new ways of storing, distributing and analysing big data about our planet. This 'digital' revolution is also accompanied by a 'sensing' revolution that is delivering unprecedented amounts of data on the state of our planet and its changes. Europe is leading this sensing revolution in space through the Copernicus initiative and the corresponding development of the Sentinel missions, which monitor our planet on an operational and sustained basis. In addition, an emerging trend, referred to as New Space in the United States or Space 4.0 in Europe, is rapidly appearing through the expanding commoditisation and commercialisation of space. In particular, the increased capabilities and the rapidly declining costs of building and launching small satellites are allowing new EO actors – including start-ups, ICT giants and other kind of actors – to enter the space business. Consequently, innovative constellations of standardised small satellites are delivering new data on our planet with high spatial resolution and high temporal frequency.

These different global data sets from space lead to a far more comprehensive picture of our planet, which is complemented by data derived from millions of smart sensors connected to the Internet (referred to as the Internet of Things, or IoT) and from unmanned aerial vehicles systems. Such streams of data offer alternative possibilities not only for science but also for entrepreneurs, who are turning big data into

[1]Head of Φ-lab Explore Office, European Space Agency (ESA), Italy

[a]Mathieu P.P., Aubrecht C., 'Earth observation open science and innovation'. Springer Nature. 2018

new types of information-based services and businesses. However, these opportunities bring new challenges for scientists, business, data and software providers seeking to exploit the vast and diverse amount of data fully by capitalising on innovative big data analytics techniques such as AI.

1.2   The Artificial Intelligence and Deep Learning revolution

AI is a general-purpose technology already transforming the global economy and many business models across industry sectors, including Space. In this context we believe AI has a huge, but still largely untapped potential for EO technology. Described as the 'new electricity' fuelling the Fourth Industrial Revolution in the 'AI for the Earth' report presented at the World Economic Forum in 2018, AI research is a real driver of innovation, future growth and competitiveness for societies and industries worldwide that can bring enormous opportunities.

Currently, AI is in the midst of a true 'renaissance', driven by Moore's law and now super-fed by big data. It is deeply entrenched in our society and routinely used in everyday life in applications ranging from recommendation engines, language services (e.g. translation, speech recognition), face recognition, virtual assistants (e.g. Siri, Alexa) and autonomous vehicles (e.g. drones, self-driving cars) – transforming the way we work and live.

What is more, over the past decade, ML has undergone a major revolution driven by the unique convergence of large-scale computing capability (e.g. cloud computing, graphics processing unit (GPU) architectures, high-performance computing, easy access to large volumes of data through the Internet and the availability of new algorithms enabling robust training of large-scale 'deep' neural networks.

Even more recently, DL and emerging reinforcement learning capabilities are becoming the 'workhorses' of AI. DL algorithms are indeed bringing dramatic improvements in the automatic recognition of objects, as first demonstrated in 2012 by results for some tasks within the ImageNet Large Scale Visualisation Challenge.

Specifically concerning EO, so far, AI applications have mainly involved computer vision to interpret and understand very high-resolution satellite imagery, but many other areas such as Earth science, prediction and big data analytics could also benefit. Therefore strengthening AI capabilities within ESA by leveraging European EO assets (including data and technical expertise) will help the science community and industry to realise the full potential of EO data in delivering socio-economic benefits and, at the same time, allow Europe to position itself in a rapidly changing AI landscape.

1.3   Rise of Artificial Intelligence for Earth Observation

AI and EO is a marriage made in heaven, however, with its own strengths as well as weaknesses. In a sense, we are now at an inflection point, at a kind of crossroads of opportunities, whereby on the one hand AI is becoming one of the most transformative technologies of the century, while on the other hand a growing European EO capability is delivering a totally unique, comprehensive and dynamic picture of the planet, thereby generating open big data sets to be explored by AI.

Due to the rapid increase in the volume and variety of EO data sources, AI techniques become increasingly necessary to analyse the data in an automatic, flexible and scalable way. Today, EO data remotely sensed from space are particularly suited – but at the same time challenging – for AI processing as they are:

- Big in size and volume with terabytes of data routinely streamed daily from space, which need to be transformed into 'small' actionable information. For example, it would take an operator several hundred years to look at the trillions of pixels routinely acquired by Sentinel-2 on a weekly basis.
- Diverse, including data from a variety of sensors, from optical (multi-spectral and hyperspectral) to radar. Up until now, AI has been mainly applied to optical imagery, in particular at very high resolution by use of traditional computer vision techniques (using mainly RGB bands). More work is needed to make full use of all available spatial, temporal and spectral information of EO data at the global scale, e.g. exploiting the full information of the complex nature of radar data within AI schemes, including information on the amplitude, frequency, phase or polarisation of the collected radar echoes.
- Complex and physically based capturing dynamic features of a highly non-linear coupled Earth system. More meaningful data extraction requires integration of physical principles into the statistical approach, and goes well beyond mere automated feature recognition where a wide variety of training data sets are available.

Machines learning (ML) algorithms powered by AI are therefore critical to needed to accelerate 'insight' into the data but should always be used in combination with domain experts vital to properly interpret the statistical correlations and data. The intersection of AI and EO remains an evolving, but a rapidly growing field. The application of ML to EO has developed rapidly over the past decade, but the emergence of deep learning (DL) has accelerated growth further, as illustrated by the increase in the number of publications. However, although very powerful, DL techniques suffer from their own inherent limitations; they are data-hungry while lacking transparency and unable to distinguish causation.

1.3.1   AI4EO data engine

An adaptative data engine for EO data based on ML is shown in Figure 1.1. It is derived from a new type of value chain, whereby the algorithms are reverse-engineered by learning from the data as opposed to the traditional way of explicitly coding knowledge.

In AI-powered value chains, most of the work is made up-front (e.g. data preparation, learning) and the result comes from the application of a trained ML scheme

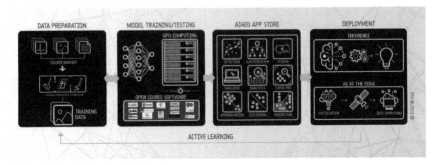

Figure 1.1   *Illustration of an adaptive data engine for EO data based on*
Machine Learning. It is derived from a new type of value chain,
whereby the algorithms are reverse engineered by learning from
the data as opposed to the traditional way of explicitly coding
knowledge. In AI-powered value chains, most of the work is made
up-front (e.g. data preparation, learning) and the result comes
from the application of a trained ML scheme (e.g. inference). The
process of transforming raw data into information becomes thereby
automatic, optimised and scalable, keeping in mind that the quality
of the output is related to the quality of the input

(e.g. inference). The process of transforming raw data into information becomes thereby automatic, optimised and scalable, keeping in mind that the quality of the output is related to the quality of the input).

1.3.2   AI@edge

Over the past decade, much of the progress in computer vision has been achieved through deep networks. Deep neural network models are computationally very demanding due to their large and growing size, already reaching several millions of parameters and neurons. Today, high-performance computers with dedicated architectures such as GPUs are needed to train and run these huge and complex models. This is why deploying DL models at the 'edge', meaning running them locally on devices such as IoT sensors, mobile phones, drones or even spacecrafts, remains a big challenge due to the limited computing power and memory of these small edge devices.

However, today, a new model to perform AI@edge is rapidly emerging driven by the amazing advances in the new generation of AI accelerator chips, which can combine high-processing power (order teraflops) with ultra-low-power consumption (1-10 W) within a very small and light footprint (e.g. supercomputer fitting into a hand). A global race for AI hardware is currently ongoing to build the most powerful, low-power and low-cost (typically below 100$) chips. This race is led by the big ICT companies such as NVIDIA with the Jetson Nano, Intel with the Myriad Vision Processing Unit and Google with Tensor Processing Unit.

Embedding the ML processing within the edge device and running it locally, i.e. bringing the computation as close as possible to the sensor, offers several advantages:

- High responsiveness. The edge computing devices can rapidly process data in real time and can take decisions autonomously without being connected. This is ideal for 'time-critical decisions' such as self-driving or for managing disasters with EO data requiring low latency for information.
- Low data rate. The AI-equipped sensor can extract in real time the relevant information from the large amount of raw data and only streams back a small amount of highly actionable data. This is ideal when there are limitations on connectivity and bandwidth for streaming data.
- Enhanced security. Processing on the sensor prevents privacy leaks, as there is less streaming of data to the cloud. This is ideal in combination with blockchain to ensure secured and automatic machine-to-machine interactions (e.g. smart contracts) within a large network of connected sensors.
- Versatility. The edge device can be easily re-programmed for new functions by uploading new parameters, through active learning or other retraining techniques. This is ideal to reconfigure the sensor or extract more capability for a suite of versatile applications, as Tesla is doing when uploading new software on the car to increase autonomy and performance.

However, running ML at the edge also presents many challenges. One key issue is the need to 'simplify' the 'complex' DL algorithms to run on edge chips. A widely used approach to address this challenge is to cut some network links and neurons to reduce complexity and memory while trying to keep the accuracy. This technique called 'pruning' is illustrated schematically in Figure 1.2. In the lab we are exploring these simplifications for a variety of EO applications.

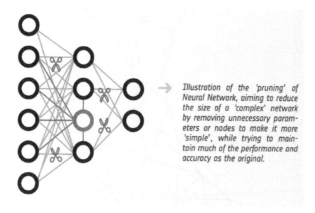

Illustration of the 'pruning' of Neural Network, aiming to reduce the size of a 'complex' network by removing unnecessary parameters or nodes to make it more 'simple', while trying to maintain much of the performance and accuracy as the original.

Figure 1.2   Illustration of the 'pruning' of neural network

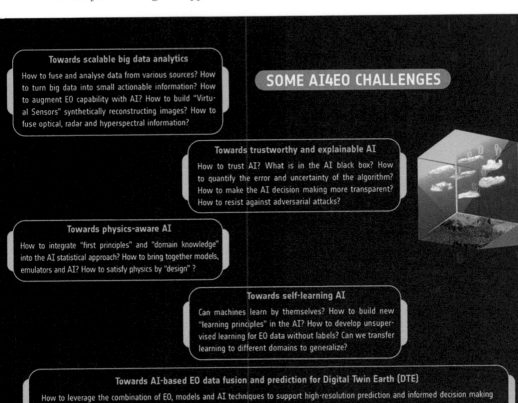

SOME AI4EO CHALLENGES

Towards scalable big data analytics

How to fuse and analyse data from various sources? How to turn big data into small actionable information? How to augment EO capability with AI? How to build "Virtual Sensors" synthetically reconstructing images? How to fuse optical, radar and hyperspectral information?

Towards trustworthy and explainable AI

How to trust AI? What is in the AI black box? How to quantify the error and uncertainty of the algorithm? How to make the AI decision making more transparent? How to resist against adversarial attacks?

Towards physics-aware AI

How to integrate "first principles" and "domain knowledge" into the AI statistical approach? How to bring together models, emulators and AI? How to satisfy physics by "design" ?

Towards self-learning AI

Can machines learn by themselves? How to build new "learning principles" in the AI? How to develop unsupervised learning for EO data without labels? Can we transfer learning to different domains to generalize?

Towards AI-based EO data fusion and prediction for Digital Twin Earth (DTE)

How to leverage the combination of EO, models and AI techniques to support high-resolution prediction and informed decision making with DTE? How to approximate the human-induced forcing and Earth System response function on our planet? How to improve prediction of the Earth System and develop "smart" satellites learning how to optimally target their observations to improve forecast?

Figure 1.3   Digital Twin Earth framework

Given this emerging high potential of AI@edge, ESA is now exploring the value to deploy AI at the ultimate edge: in Space. In this context, ESA has launched (September 3, 2020) a CubeSat mission called FSSCat including a suite of sensors but also an experiment dedicated to AI on-board. The idea of the so-called Φ-sat-1 mission is to run AI on the Intel Myriad-2 chip to identify the relevant data of the Hyperscout-2 sensor to download (meaning without cloud cover).

This experiment could open new avenues for innovative use of AI for space and AI in space, whereby the sensors become 'smarter' as they capture the 'meaning' of the image while capturing light. It could also explore new ways to upgrade the sensor capabilities by re-programming the brain of the satellite for versatile applications. This experiment is just the beginning of a family of Φ-sat missions. We are only scratching the surface of what AI can do for space and how it can shape it in the next decade.

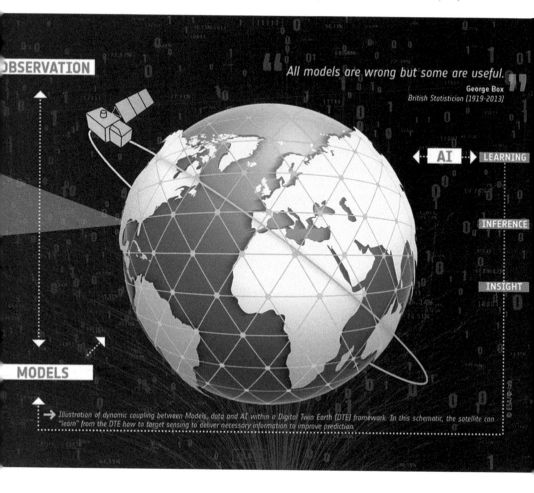

OBSERVATION

All models are wrong but some are useful.

George Box
British Statistician (1919-2013)

AI → LEARNING

INFERENCE

INSIGHT

MODELS

© ESA/@-lab

→ Illustration of dynamic coupling between Models, data and AI within a Digital Twin Earth (DTE) framework. In this schematic, the satellite can "learn" from the DTE how to target sensing to deliver necessary information to improve prediction.

1.4   Conclusions

There is a pressing need to increase Europe's strength and positioning in the area of AI research applied to space. In particular, EO big data presents a niche area of the data economy. Much more could be gained from these valuable data by using AI, thus increasing opportunities for a growing variety of applications and services. The investment, research and innovation efforts required to achieve such an ambitious goal are huge, and cannot be achieved alone, but only through teaming and partnership. European Space Agency (ESA) can play an important role in such endeavours along with other partners at the European level.

Today, the power of AI for EO is largely untapped. Many challenges still need to be tackled at scientific, applications and capability levels to deliver the maximum value from open EO data from satellites for our society and economy. Some challenges are presented as follows:

- **Frontier scientific problems** – including not only fundamental and applied research, ranging from classification, detection, indexing, understanding, prediction, data fusion, inversion problems, up to automation, but also new generation of applications and services.
- **Bringing AI at the edge** – including the development of autonomous systems performing AI inference on board as well as intelligent swarm of satellites powered by AI and guided by machine-to-machine communication.
- **Digital Twin Earth (DTE)** – integrating various data sources (including EO), models (Earth simulations) and AI to develop an interactive 'digital replica' of the entire planet, which would provide scientists and decision-makers a unique tool to quantify our impact on the planet and check how to adjust our trajectory towards a more sustainable future (see Figure 1.3).
- **Building capacity** – ensuring that Europe supports the talents, skills and resources to develop AI4EO.

It is critical to address these challenges at the European level, by leveraging ESA assets, and fostering partnership, to realise the full potential of EO and deliver its full socio-economic benefits. EO is capable of delivering a unique and comprehensive data set about the planet which can be analysed using AI, in ways that are mutually transformative of both EO and AI. In that sense, EO data with its unique and complex features could also help drive some advances in AI techniques, by trying to make the full use of all spectral, temporal and spatial EO information.

In this chapter the key concepts of AI4EO have been introduced, with particular focus on the enormous innovation that these technologies are able to introduce in the world of research.

This book is structured as follows: it starts from the basics of remote sensing, and introduces the theoretical and mathematical concepts related to AI, ML and DL. Subsequently, the mechanisms for creating data sets are introduced and some AI4EO applications are presented. The book concludes with a chapter on AI@edge.

Chapter 2

Principles of satellite data analysis

Alessandro Sebastianelli[1], Silvia Liberata Ullo[1], Maria Pia Del Rosso[1], Sara Aparício[2], Andrea Radius[3], Chiara Zarro[1], James Wheeler[4], Chris Stewart[4], and Pia Addabbo[5]

This chapter introduces the principles underlying satellite remote sensing, highlighting the analysis and manipulation of data acquired by the related platforms. In particular, the satellites of the ESA (European Space Agency) Copernicus programme are taken as a reference for a detailed description of the chapter, because of the ESA policy to make Copernicus data freely available. As a consequence of this choice, several case studies will be presented in the other chapters of the book, where data acquired by the Sentinels of the Copernicus programme are mainly used.

2.1 Introduction

Remote sensing is a way of collecting and analysing data to get information about an object without a direct contact with the object itself. In remote sensing, four elements are essential:

- the *instrument* or sensor on board of the platform
- the *platform* that mounts the instrument
- the *object* (target) to be observed
- the *information* obtained from the acquired data, how it is stored and used

When Earth scientists talk about remote sensing, the observed object is the Earth. In general terms then, remote sensing is a tool to observe and study the Earth, its interior, its land surface, the oceans, the atmosphere and its dynamics, from space. Moreover, the platforms are all the means used to be 'at a distance' from the

[1]Department of Engineering, University of Sannio, Italy
[2]Solenix/ESA, Italy
[3]ICEYE, Finland
[4]Φ-Lab, European Space Agency (ESA), Italy
[5]University Giustino Fortunato, Italy

Earth's surface, such as planes and satellites, aircrafts and even drones. The target is our planet itself, the sensors are contained in all the instruments used to observe the Earth (cameras, scanners, radars, etc.) and the information obtained is everything that increases our knowledge about our planet, such as cloud cover, the evolution of the ozone hole, the spreading of the deserts, the progress of deforestation and much more [1].

2.1.1   *Principles underlying remote sensing*

Remote sensing means measuring an object remotely, without any physical contact with the object itself. Vision is also a form of remote sensing, because when the eye looks at an object, it records the electromagnetic radiation (reflected light) coming from the surface of that object, while illuminated by external light sources. The electromagnetic radiation contains information on the surface of the object itself and allows to detect its characteristics, such as its shape and the colour our brain associates to it. It is important to note that the human eye can perceive only radiation within a small area of the electromagnetic spectrum. So the tools for remote sensing outside the visible wavelengths represent an extension of our field of vision and allow us to get additional information on the external world [2–5]. Moreover, this type of instruments not only allows a qualitative analysis, but it provides sector specialists to quantitatively measure a certain number of parameters that characterise our planet.

2.1.1.1   **Electromagnetic radiation**

An electromagnetic wave is generated by the oscillation of a magnetic field and an electric field, perpendicular to the direction of propagation (see Figure 2.1).

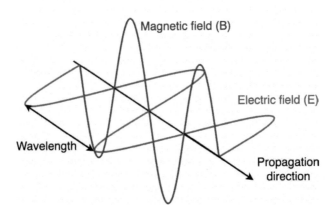

Figure 2.1   Electromagnetic wave

Regardless of the vibration frequency of the fields, electromagnetic radiation propagates in empty space at the speed of light c.

For simplicity we can describe an electromagnetic wave through four parameters:

- Wavelength λ: It is the distance from one wave crest to the next. Wavelength can be measured in metres.
- Frequency f: It is measured as the number of crests passing a fixed point in a given period of time. Frequency is often measured in hertz, equivalent to cycles per second, and its multiples.
- Amplitude A: It is equivalent to the height of each peak. Amplitude is often measured as energy levels (also known as spectral radiance).
- Phase: The phase of a waveform specifies the extent to which the peaks of one waveform align with those of another. Phase is measured in angular units, such as degrees or radians. If two waves are aligned, they oscillate together and are said to be 'in phase' (a phase shift of 0 degrees). However, if a pair of waveforms are aligned such that the crests of one waveform match with the minimum values of the other, they are said to be 'out of phase' (a phase shift of 180 degrees).

The propagation speed is constant and is given by the product of the wavelength and the frequency. Therefore, high wavelengths correspond to low frequencies and vice versa. Light consists of electromagnetic waves, which are visible to the human eye when their wavelength is between about 0.38 and 0.74 (visible band), and it determines the distinction of the colours.

2.1.1.2   Electromagnetic spectrum

Depending on their frequency, electromagnetic waves are produced by different types of sources, have different properties and different ways of interacting with the objects.

The frequency intervals most often used in remote sensing are:

- *Visible*: Restricted part of the spectrum, perceptible by the human eye, allows us to distinguish objects and their colours; includes wavelengths from about 0.38 µm (corresponding to the colour purple) to 0.74 µm (corresponding to the colour red).
- *Infrared*: It consists of wavelengths greater than those of the visible, ranging between 0.7 µm and 1 mm. It is divided into near infrared, medium infrared, far infrared or thermal, extreme infrared. This type of radiation is linked to interaction phenomena between energy and matter and is capable of giving information about the surface temperature of objects.
- *Microwave*: The wavelength varies from 1 mm to 1 m, and these waves are widely used for both active and passive remote sensing. They allow to obtain different types of information, such as surface roughness or their water content.

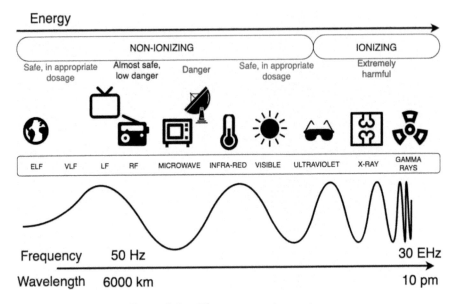

Figure 2.2   Electromagnetic spectrum

By ordering the electromagnetic energy according to the wavelength, or equivalently according to the frequency, the electromagnetic spectrum is obtained (see Figure 2.2), which for practical convenience is divided into spectral bands.

2.1.2   *Interaction between electromagnetic waves and matter*

In general, a ray incident on an obstacle, or on the separation surface between two different media, generates reflection, refraction and diffraction phenomena. However, there are additional mechanisms by which an electromagnetic wave can interact with matter.

2.1.2.1   Reflection

Reflection is the phenomenon whereby an electromagnetic wave hits a separation surface between two media, and continues its path in the same middle of departure, as shown in Figure 2.3. Most commonly we speak of total reflection, and all the incident electromagnetic radiation is sent forward with respect to the source, in this case. According to the Snell's law, and as a function of the refractive indices of the two media, n_1 and n_2, it is possible that a part of the electromagnetic wave crosses the boundary between the two media, by giving rise to the phenomenon of diffraction, discussed onwards [6].

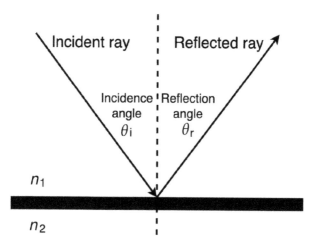

Figure 2.3   Reflection

Total internal reflection law

By defining $\theta_1 = \theta_i$ the angle of incidence of the light beam and defining θ_r the angle formed by the reflected ray with the normal to the surface, we have that $\theta_i = \theta_r$. If instead we call $\theta_2 = \theta_t$ the angle formed by the ray refracted (see Figure 2.4), according to Snell's law, we have that $n_1 \sin \theta_i = n_2 \sin \theta_t$, with n_1 and n_2 being the refractive indices of the media.

Total internal reflection occurs when the angle θ_t reaches the amplitude of $\frac{\pi}{2}$, that is the case when there is no longer a refracted wave. This phenomenon can occur in the passage from a denser medium to a less dense one ($n_1 > n_2$); the angle θ_i is called critical angle and is given by

$$\theta_{\text{critic}} = \arcsin \frac{n_2}{n_1} \tag{2.1}$$

When $\theta_i > \theta_{\text{critic}}$, no refracted beam appears: the incident light undergoes a total internal reflection by the interface. A surface wave is generated, also called leaky wave, which decays exponentially within the medium with refractive index n_2.

2.1.2.2   Refraction

Refraction is the change in direction of a wave passing from one medium to another as shown in Figure 2.4. The light refraction is the most commonly observed phenomenon, but other waves such as sound waves and water waves also experience refraction. How much a wave is refracted depends on the change in the electromagnetic

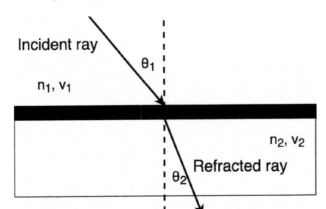

Figure 2.4   Refraction

wave speed and on the initial direction of the wave propagation. For the light, refraction follows the Snell's law, as anticipated before.

Snell's law

For a given pair of media, the ratio of the sines of the angle of incidence θ_1 and angle of refraction θ_2 is equal to the ratio of phase velocities (v_1 / v_2) in the two media, or equivalently to the indices of refraction (n_2 / n_1) of the two media:

$$\frac{\sin \theta_1}{\sin \theta_2} = \frac{v_1}{v_2} = \frac{n_2}{n_1} \tag{2.2}$$

When one of the two media is the vacuum, the absolute refractive index is defined as the ratio between the speed of light and the speed of propagation in the generic medium [6].

2.1.2.3   Diffraction

Diffraction occurs when a wave encounters an obstacle of a size comparable to the wavelength of the electromagnetic wave, or when an aperture is present on the illuminated surface, as represented in Figure 2.5. In this case, the wave also propagates in the part that should be in the shade, bypassing the obstacle. The wave that propagates is no longer flat, but it bends over the edges [6].

2.1.3   Other types of interaction

The Doppler effect occurs when the source, target or propagation medium are in motion (see Figure 2.6). If the electromagnetic wave source moves towards the

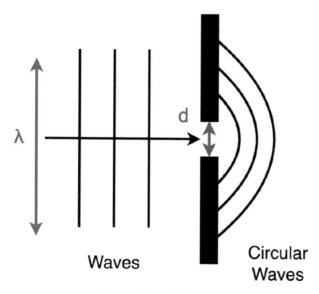

Figure 2.5   Diffraction

target, each subsequent wave crest is emitted in a position closer to the target with respect to the previous one, at the same frequency. So each wave will take less time to reach the target than the previous one. Yet, for the target, the time interval between the reception of each wave crest will decrease, generating an increase in the perceived frequency. By measuring the reflected wave, the frequency variation will be double compared to the effect on the incident wave. The total effect is obtained by analysing the movement of the source, the target and the medium separately [7, 8].

Transmission of radiation occurs when radiation passes through a substance without a significant attenuation. From a given depth of a substance, the ability of a medium to transmit energy is measured as the transmittance, expressed by the following equation:

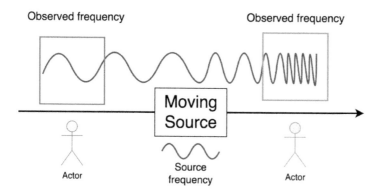

Figure 2.6   Doppler effect

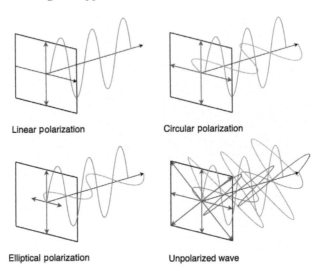

Linear polarization   Circular polarization

Elliptical polarization   Unpolarized wave

Figure 2.7   Different types of polarisation

$$t = \frac{\text{transmitted radiation}}{\text{incident radiation}} \qquad (2.3)$$

In the field of remote sensing, the transmittance of films and filters is often important. With respect to naturally occurring materials, we often think only of water bodies, as capable of transmitting significant amounts of radiation. However, the transmittance of many materials varies greatly with the wavelength. For instance, the plant leaves are generally opaque to visible radiation but transmit significant amounts of radiation in the infrared. Therefore, the transmittance of a material depends on the wavelength.

Fluorescence occurs when an object illuminated with an electromagnetic radiation at a specific wavelength emits radiation at a different wavelength.

The *polarisation* of electromagnetic radiation denotes the orientation of the oscillations within the electric field of electromagnetic energy (see Figure 2.7). A light wave's electric field (travelling in a vacuum) is typically oriented perpendicular to the wave's direction of travel (i.e. the energy propagates as a transverse wave); the field may have a preferred orientation, or it may rotate as the wave travels. Polarisation has broad significance in the practice of remote sensing, because, for example, within the atmosphere, polarisation of light is related to the nature and abundance of atmospheric aerosols and atmospheric clarity [3–5, 9].

2.1.4   Interaction with the atmosphere

All radiation used for satellite remote sensing must pass through the Earth's atmosphere. If the sensor is mounted on a low-flying platform, the effects of the atmosphere on the electromagnetic radiation are negligible. On the contrary, if sensors mounted on satellites are considered, the electromagnetic radiation must pass through the entire depth of the Earth's atmosphere and so the atmospheric effects have a greater impact on the quality of the acquired data. Therefore, remote sensing experts must have the right knowledge of the interactions between the electromagnetic radiation with the atmosphere. The most common physical processes due to the presence of the atmosphere are scattering, absorption and refraction [3–5, 9].

2.1.4.1   Scattering

Scattering is the redirection of electromagnetic energy by particles suspended in the atmosphere. The amount of scattering that occurs depends on the sizes of these particles, their abundance, the wavelength of the radiation and the portion of the atmosphere that the radiation have to pass through. Through the scattering phenomenon, a part of the transmitted radiation is directed back towards the source and a part towards the Earth's surface. Common forms of scattering are:

- *Rayleigh scattering*: It occurs when atmospheric particles have very small diameters compared to the wavelength of the radiation. This type of scattering is the reason for the blue colour of the sky and the red and orange colours of the sunset. It is the dominant scattering process at high levels in the atmosphere (up to altitudes of 9–10 km). Rayleigh scattering is wavelength-dependent, meaning that the amount of scattering changes greatly as one examines different regions of the spectrum.
- *Mie scattering*: It is caused by large atmospheric particles, including dust, pollen, smoke and water droplets. Those particles have diameters comparable with the wavelength of the scattered radiation. Mie scattering can influence a broad range of wavelengths in and near the visible spectrum. Mie scattering is wavelength-dependent, but not in the simple manner of Rayleigh scattering; it tends to be greatest in the lower atmosphere (0–5 km), where larger particles are abundant.
- *Nonselective scattering*: It is caused by particles that are much larger than the wavelength of the scattered radiation. This type of scattering is not wavelength-dependent.

For remote sensing, scattering has several important consequences. Because of the wavelength dependency of the scattering, in almost all its forms, radiation in the blue and ultraviolet regions of the spectrum (which are most strongly affected by scattering) is usually not considered useful for remote sensing. Images that record these portions of the spectrum tend to record the brightness of the atmosphere rather than the brightness of the scene itself. Scattering tends to make dark objects appear brighter than they would otherwise be, and bright objects appear darker, thereby decreasing the contrast recorded by a sensor. Because 'good' images must preserve the range of brightness present in a scene, scattering would degrade the quality of an image [3–5, 9], in those regions of the spectrum.

2.1.4.2   Absorption

Absorption is a phenomenon occurring to radiation when its transmission through the atmosphere is attenuated or completely unabled. The absorption depends on the concentration of gases in the atmosphere. The presence of ozone (O_3) and oxygen (O_2) prevents the transmission of short-wavelength radiation (ultraviolet band) from the upper atmospheric layer to the lower one. In the lower atmosphere, the same function is done by carbon dioxide (CO_2), mostly uniform except for local variations due to human activities and volcanic eruptions. This gas is responsible for absorption in the mid- and far infrared bands [3–5, 9].

2.1.4.3   Refraction

Another phenomenon caused by atmosphere is refraction: the passage through atmospheric layers with different values of humidity, temperature and clarity make the light rays blend in different directions. This is the same phenomenon that brings us to see far object shimmering near the surface of roads and parking lots with hot air [3–5, 9].

2.2   Types of sensors

The instruments mounted on board a remote sensing platform can be divided into two main categories: active and passive. Active sensors provide their own source of energy to 'illuminate' the objects they observe. An active sensor emits radiation in the direction of the target to be investigated, and usually it detects and measures the radiation that is reflected or backscattered from the target (see Figure 2.8). Passive sensors, on the other hand, detect the energy that is emitted or reflected by the object or the scene being observed, without radiating the surface of interest, hence the term 'passive sensors' (see Figure 2.9). Reflected sunlight is the most common source of radiation measured by passive sensors. Nowadays, the lights emitted during the night can also be a source of radiation measured through the passive sensors on board of the satellites.

2.2.1   Active sensors

The majority of active sensors operate in the microwave portion of the electromagnetic spectrum, which makes them able to penetrate the atmosphere under most conditions. Yet, some active sensors work also in the optical frequency interval.

An active technique views the target from either end of a baseline of known length. The change in apparent view direction (parallax) is related to the absolute distance between the instrument and target. Active remote sensors include the following:

- *LiDAR*: A *l*ight *d*etection *a*nd *r*anging sensor that uses a laser radar to transmit a light pulse and a receiver with sensitive detectors to measure the backscattered or reflected light. Distance to the object is determined by recording the time

Figure 2.8   Example of active sensor

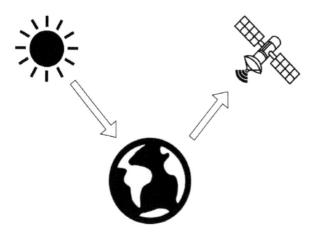

Figure 2.9   Example of passive sensor

between transmitted and backscattered pulses and by using the speed of light to calculate the distance travelled.

• *Laser altimeter*: An instrument that uses a LiDAR to measure the height of the platform (spacecraft or aircraft) above the surface. The height of the platform with respect to the mean Earth's surface is used to determine the topography of the underlying surface.

- *RADAR*: An active *ra*dio *d*etection *a*nd *r*anging sensor that provides its own source of electromagnetic energy. An active radar sensor, whether airborne or spaceborne, emits microwave radiation in a series of pulses from an antenna. When the energy reaches the target, some of the energy is reflected back towards the sensor. This backscattered microwave radiation is detected, measured and timed. The time required for the radiation to travel to the target and return back to the sensor determines the distance or range to the target.

- *Ranging instrument*: A device that measures the distance between the instrument and a target object. Radars and altimeters work by determining the time a transmitted pulse (microwaves or light) takes to reflect from a target and return to the instrument. In some cases, identical microwave instruments are on board of a pair of platforms. Signals are transmitted from each instrument to the other, with the distance between the two determined through the calculation of the difference between the received signal phase and transmitted signal phase, taken as a reference.

- *Scatterometer*: A high-frequency microwave radar designed specifically to measure backscattered radiation. Over ocean surfaces, measurements of backscattered radiation in the microwave spectral region can be used to derive maps of surface wind speed and direction.

- *Sounder*: An instrument that measures vertical distribution of precipitation and other atmospheric characteristics such as temperature, humidity and cloud composition.

In Figure 2.6 an example of an active sensor on board of a satellite is shown.

2.2.2 Passive sensors

Passive sensors include different types of radiometers and spectrometers.

Most passive systems used in remote sensing applications operate in the visible, infrared, thermal infrared and microwave portions of the electromagnetic spectrum. In Figure 2.7 an example of a passive sensor on board of a satellite is shown. Passive remote sensors include the following:

- *Accelerometer*: An instrument that measures acceleration (change in velocity per time unit). There are two general types of accelerometers. One measures translational accelerations (changes in linear motions in one or more dimensions), and the other measures angular accelerations (changes in rotation rate per time unit).

- *Multispectral sensor*: It measures the energy reflected within several specific sections – or *bands*, of the electromagnetic spectrum, like visible and near-infrared (VNIR). Multispectral sensors usually have between 3 and 15 different band measurements in each pixel of the images they produce.

- *Hyperspectral radiometer*: An advanced multispectral sensor that detects hundreds of very narrow spectral bands throughout the visible, near-infrared and mid-infrared portions of the electromagnetic spectrum. This sensor has a very high spectral resolution, which facilitates fine discrimination between different targets based on their spectral response in each of the narrow bands. The main difference between multispectral and hyperspectral is the number of bands and how narrow the bands are.
- *Radiometer*: An instrument that quantitatively measures the intensity of electromagnetic radiation in some bands within the spectrum. Usually, a radiometer is further identified by the portion of the spectrum it covers, e.g. visible, infrared or microwave.
- *Imaging radiometer*: A radiometer that has a scanning capability to provide a two-dimensional (2D) array of pixels from which an image may be produced. Scanning can be performed mechanically or electronically by using an array of detectors.
- *Sounder*: An instrument that measures vertical distributions of atmospheric parameters such as temperature, pressure and composition from multispectral information.
- *Spectrometer*: A device that is designed to detect, measure and analyse the spectral content of incident electromagnetic radiation. Conventional imaging spectrometers use gratings or prisms to disperse the radiation for spectral discrimination.
- *Spectroradiometer*: A radiometer that measures the intensity of radiation in multiple wavelength bands (i.e. multispectral). Many times the bands are of high spectral resolution, designed for remotely sensing specific geophysical parameters.

2.3 Platforms

Depending on the altitude, that is the height at which a platform is located with respect to the Earth's surface, the platforms are classified into three types (see Figure 2.10).

2.3.1 Ground-based platforms

These platforms are primarily located on the ground. Some of these platforms are placed at a certain height, however they are directly supported from the ground. Hence, ground platforms can be grouped into platforms which are present on the ground or at a low height from the ground, i.e. in the order of a few metres. Examples of these platforms include handheld cameras, and cameras mounted on vehicles, towers, cranes, etc.

Sensors from all these platforms capture data from very low heights. As the height of photography is very low, the captured data is normally of very high resolution. Smaller areas are covered by a single capture. Hence, to study a large area many photographs are required. These photographs are conventionally taken as reference data or ground truth and are mostly captured as sample data. This data is very simple to understand as it directly represents the objects on the ground. More time is required for capturing a large city. However, complete control of photography is possible as it is performed from the ground itself. Most often photography or remote sensing is performed in visible spectra in this case.

2.3.2   *Airborne platforms*

These are platforms which are present at a relatively higher altitude from the ground surface. As the name implies, they are practically present in the air at the time of photography, i.e. data capture. The height of platforms used for photography or remote sensing ranges from a few metres (in the case of drones) to a few kilometres (in the case of aeroplanes). Examples of airborne platforms include aerial cameras and sensors mounted on aeroplanes, helicopters, drones and hot air balloons.

As platform altitude is higher than ground-based platforms, here more area coverage is possible in a single instance. In less time and with fewer photographs, it is possible to cover an entire city. The photographs captured can be understood by the general public, although in some instances there will be some difficulties depending

Ground-based platforms

Ground-based platforms include:
- hand held cameras (film or digital)
- cranes
- ground vehicles
- tethered balloons
- towers.

Airborne platforms

Airborne platforms include:
- airplanes
- helicopters
- unmanned aerial systems (UAS)
- high-altitude aircraft
- free-floating balloons.

Spaceborne platforms

Spaceborne platforms include:
- the space station
- low-level (700 to 1 500 km) satellites
- high-level (~ 36 000 km) satellites.

Figure 2.10   Remote sensing platform

on the resolution of an image (relating to the size of objects appearing in photos). Multispectral sensors can also be used to perform remote sensing in this case.

The applications of airborne platform-based remote sensing include reconnaissance surveys, disaster management and aerial surveys. However, it is not possible to totally control the remote sensing process, as the sensors are held in the air. It is practically not possible to maintain complete uniform velocity and smooth propagation of an aeroplane (platform) during the photography. Due to wind and platform perturbations, some amount of distortions are observed in the aerial photographs. The distortions include crab and drift.

2.3.3 Spaceborne platforms

These platforms are present at a great height from the earth's surface. The altitude of platforms range from a few hundred kilometres to several thousand kilometres. Examples of these platforms include satellites, space shuttles and rockets. Spaceborne platforms travel around the Earth following different orbits. Polar orbiting satellites revolve around the Earth in an orbit ranging from a few hundred kilometres to a few thousand kilometres, whereas geosynchronous satellites used in remote sensing orbit at a height of approximately 36 000 km. These satellites capture data continuously while moving in their orbits around the Earth and transmit it to ground stations. Almost all sensors listed in Section 2.2 are or have been carried on spaceborne platforms. Also, depending on the sensor used and resolution characteristics of the sensor, the data captured are used in a wide range of applications. These include weather, climate, forest and agriculture mapping, glacier studies and urban studies.

A large area can be captured in a single scene (or photo) depending on altitude and resolution of the sensor. However, the process cannot be controlled completely from the ground. For reasons such as the platform velocity, atmospheric variations and sensor problems, a lot of noise is included in the data collected. To remove noise, different types of image processing operations are performed at the ground level. It is very difficult for novices and members of the general public to understand all details present in the image captured using spaceborne sensors.

2.4 Data characteristics

The quality of remote sensing data also depends on a particular characteristic, called resolution, which is represented by four types of different parameters: spatial, spectral, radiometric and temporal resolution [3–5].

Spatial resolution: It represents the size of the smallest possible feature that can be detected (see Figure 2.11). The spatial resolution quantifies the capability to separate two close targets. The pixel size is often considered as spatial resolution; for clarity, the spatial resolution depends on the design of the sensor, while the pixel size depends on the digital sampling of the signal. However, the pixel size can be considered quantitatively related to the spatial resolution, because it is defined imposing constraints depending on the spatial resolution. The size of a pixel that

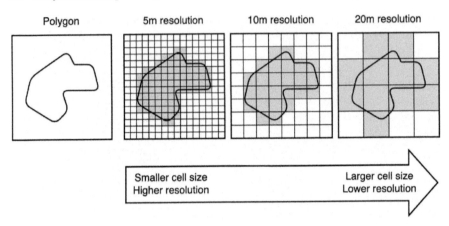

Figure 2.11   Spatial resolution

is recorded in a raster image may correspond to square areas ranging in side length from a few decimetres to thousands of metres.

Spectral resolution: It represents the wavelength of the different frequency bands recorded usually, and this is related to the number of frequency bands recorded by the platform. Spectral resolution describes the ability of a sensor to define fine wavelength intervals. The finer the spectral resolution, the narrower the wavelength range for a particular channel or band (see Figure 2.12).

Radiometric resolution: It refers to the number of different intensities of radiation the sensor is able to distinguish. Radiometric resolution is the capacity of the instrument to distinguish differences in light intensity or reflectance (see Figure 2.13). The greater the radiometric resolution, the more accurate the sensed image will be [10]. Typically, this ranges from 8 to 14 bits, corresponding to 256 levels of the greyscale and up to 16 384 intensities or 'shades' of colour, in each band. It has affected by the instrument noise.

Temporal resolution: It is the frequency of flyovers by the satellite or plane. This resolution can become relevant in time series studies or those requiring an

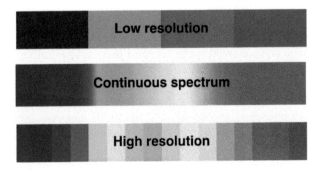

Figure 2.12   Spectral resolution

8-bit radiance level

2-bit radiance level

Figure 2.13   Radiometric resolution

averaged or mosaic image, such as for deforestation monitoring. Moreover, temporal resolution plays an important role when data and information related to them are needed in almost real time (i.e. forest monitoring for fire detection, for instance).

2.4.1   Operations for improving quality of data

In order to create sensor-based maps, most remote sensing systems expect to calibrate sensor data in relation to a reference point including distances between known points on the ground. This depends on the type of sensor used. For example, in conventional photographs, distances are accurate in the centre of the image, with the distortion of measurements increasing the farther you get from the centre. Another factor is that of the platen against which the film is pressed can cause severe errors when photographs are used to measure ground distances. The step in which this problem is resolved is called georeferencing, and involves computer-aided matching of points in the image (typically 30 or more points per image) which is extrapolated with the use of an established benchmark, 'warping' the image to produce accurate spatial data [3–5]. In addition, images may need to be radiometrically and atmospherically corrected:

- *Radiometric correction*: It allows avoidance of radiometric errors and distortions. The illumination of objects on the Earth surface is uneven because of different properties of the relief. This factor is taken into account in the method of radiometric distortion correction. Radiometric correction gives a scale to the pixel values, e.g. the monochromatic scale of 0–255 will be converted to actual radiance values.
- *Topographic correction*: In rugged mountains, as a result of terrain, the effective illumination of pixels varies considerably. In a remote sensing image, the pixel on the shady slope receives weak illumination and has a low radiance value, in contrast, the pixel on the sunny slope receives strong illumination and has a high radiance value. For the same object, the pixel radiance value on the shady slope will be different from that on the sunny slope. Additionally, different objects may have similar radiance values. These ambiguities seriously affected remote sensing image information extraction accuracy in mountainous areas. It became the main obstacle to further application of remote sensing

images. The purpose of topographic correction is to eliminate this effect, recovering the true reflectivity or radiance of objects in horizontal conditions. It is the premise of quantitative remote sensing application.

- *Atmospheric correction*: Elimination of atmospheric haze by rescaling each frequency band so that its minimum value (usually realised in water bodies) corresponds to a pixel value of 0. The digitising of data also makes it possible to manipulate the data by changing greyscale values.

2.5   Copernicus programme

Copernicus is an Earth observation programme – and the biggest provider of Earth observation data in the world. It provides accurate, timely and easily accessible information to improve the management of the environment, understand and mitigate the effects of climate change and ensure civil security. Copernicus is the new name for the Global Monitoring for Environment and Security programme, previously known as GMES. This initiative is headed by the European Commission in partnership with the ESA [11].

2.5.1   The Sentinels

ESA is developing a new family of satellites, called Sentinels, specifically for the operational needs of the Copernicus programme. At present, three complete two-satellite constellations are in orbit plus an additional single satellite, Sentinel-5P.

Each Sentinel mission is based on a constellation of two satellites to fulfil revisit and coverage requirements, providing robust datasets for Copernicus services. Copernicus provides a unified system through which vast amounts of data are fed into a range of thematic information services designed to benefit the environment, the way we live, humanitarian needs and support effective policymaking for a more sustainable future. These services fall into six main categories: land management, the marine environment, atmosphere, emergency response, security and climate change. The Sentinels carry a range of technologies, such as radar and multispectral imaging instruments for land, ocean and atmospheric monitoring:

- *Sentinel-1*: It is a polar orbiting satellite which provides all-weather, day and night radar imagery for land and ocean services. Sentinel-1A and -1B were launched, respectively, in April 2014 and April 2016. Both were taken into orbit on a Soyuz rocket from Europe's Spaceport in French Guiana. This mission carries on board a synthetic aperture radar (SAR) operating at C-band (central frequency of 5.404 GHz).
- *Sentinel-2*: It is a polar orbiting, multispectral high-resolution imaging mission for land services. It provides, for example, imagery of vegetation, soil and water cover, inland waterways and coastal areas. Sentinel-2 can also deliver information for emergency services. Sentinel-2A was launched on 23 June 2015 and

Sentinel-2B followed on 7 March 2017. Each Sentinel-2 carries a single multi-spectral instrument (MSI) sampling 13 spectral bands.

- *Sentinel-3*: It is a multi-instrument mission which provides high-accuracy optical, radar and altimetry data for marine and land services, to measure sea surface topography, sea and land surface temperature, ocean colour and land colour with high-end accuracy and reliability. The mission will support ocean forecasting systems, as well as environmental and climate monitoring. The two Sentinel-3 satellites were launched on 16 February 2016 and 25 April 2018. Sentinel-3 carries a spectrometer instrument called Ocean and Land Colour Instrument (OLCI), an imaging radiometer named Sea and Land Surface Temperature Radiometer (SLSTR) and a Microwave Radiometer (MWR).
- *Sentinel-4*: It is a payload devoted to atmospheric monitoring that will be embarked upon a Meteosat Third Generation-Sounder (MTG-S) satellite in geostationary orbit.
- *Sentinel-5 Precursor* (also known as Sentinel-5P): It is the forerunner of Sentinel-5 to provide timely data on a multitude of trace gases and aerosols affecting air quality and climate. It has been developed to reduce data gaps between the Envisat satellite, in particular the Sciamachy instrument, and Sentinel-5. Sentinel-5P, the first Copernicus mission in orbit dedicated to monitoring our atmosphere, was taken into orbit on a Rockot launcher from the Plesetsk Cosmodrome in northern Russia on 13 October 2017. Sentinel-5P carries a spectrometer named the TROPOspheric Monitoring Instrument (TROPOMI).
- *Sentinel-5*: It is a payload that will monitor the atmosphere from polar orbit aboard a MetOp Second Generation satellite.
- *Sentinel-6*: It carries a radar altimeter to measure global sea surface height, primarily for operational oceanography and for climate studies.

2.6 Focus on optical and SAR data

In this section we will focus on Sentinel-1 (SAR) and Sentinel-2 (Optical) data, because they encompass the most commonly used remote sensing platforms, but also because the applications that will be presented in the next chapters are mainly focused on the use of these types of data. So in the next subsections the reader will find the most useful notions related to SAR and Optical data, their principles and the most common pre-processing stages and techniques. Obviously these data are not used only in the field of artificial intelligence for Earth observation, but each of them can be used for various applications [12–25], such as the measurement of the vegetation index (normalised difference vegetation index, NDVI), the mapping of water basins, the measurement of displacements following seismic events, mapping of flooding and landslides, etc.

2.6.1 Principles of SAR

SAR is a radar imaging technique which synthesises a larger aperture to improve spatial resolution in the azimuth direction. Imaging radar systems measure the return signal from a sequence of pulses emitted from a side looking geometry, perpendicular to the line of flight. Sentinel-1 is a typical example of a satellite with the SAR imaging technique.

The resolution in range (perpendicular to flight path) of a radar imaging system depends on the ability to distinguish the return pulses from two targets. The range resolution would theoretically improve with a shorter pulse. However, the shorter the pulse, the weaker the power that can be generated. To avoid reducing the pulse length, while maximising range resolution, a frequency modulated pulse is used. The range resolution then depends on the pulse beam width.

SAR – basic principles

A SAR is an imaging radar mounted on a moving platform. Electromagnetic waves are transmitted sequentially, the echoes are collected and the system electronics digitises and stores the data for subsequent processing. As transmission and reception occur at different times, they map to different positions. The well-ordered combination of the received signals builds a virtual aperture that is much longer than the physical antenna width, i.e. the source of the term 'synthetic aperture', giving it the property of an imaging radar.

The azimuth (the direction along the flight path) resolution of a real aperture radar depends on the width of the beam projected on the ground. This depends on the length of the antenna and the distance to the target. To achieve a high azimuth resolution, a very large antenna would be required, which is not feasible for spaceborne platforms. An alternative to improve the azimuth resolution is achieved through aperture synthesis.

SAR improves the azimuth spatial resolution by exploiting the Doppler frequency shift, caused by the change in relative velocities between the ground and the instrument as it orbits the Earth. The length of this 'synthetic aperture' is approximately equal to the beam width provided by the actual antenna. Pulses are transmitted from small antennas with large overlapping footprints as they move in azimuth. Azimuth compression uses the phase histories of objects on the ground from pulses emitted in different positions along the track to greatly improve the azimuth spatial resolution [26, 27]. Raw SAR data, including that from Sentinel-1, are typically processed into two different levels of distribution to the users:

- *Single look complex (SLC) data*: In SLC images, each pixel contains both the phase and amplitude of the backscatter with a single look and in slant range geometry. In single look data, the azimuth compression has been carried out using the full azimuth bandwidth of the sensor. This produces the highest azimuth spatial resolution, but at the cost of higher speckle. The image pixels are usually not square, given that the range and azimuth resolutions are not the same. SLC images are required for applications that use the phase, such as interferometry.

- *Multilook detected (MLD) or ground range detected (GRD)*: For applications that do not require the phase, it is often more convenient to use MLD images. MLD images have been multilooked, usually to produce pixels of equal dimensions in range and azimuth. As the multilook process is done by incoherent averaging, the phase information is lost. This type of data processing will be discussed later.

2.6.2 Classic processing workflows for SAR data

The data received from the antenna are unfocused and need to be thoroughly processed to provide the desired information on the backscattered energy. The SAR processing compresses the data through a 2D correlation, which allows focusing in the range and azimuth direction. The resolution of the 2D correlation is not possible in closed-form and must be approximated mathematically to allow focusing. Many SAR processing algorithms have been proposed [28]; here, for simplicity, we can neglect the details of the SAR processing, as it is not the aim of this book, and assume that the signal is properly focused in both directions after the processing.

The SAR products can be classified into three levels. The Level 0 (L0) products consist of raw unfocused data, which need subsequent SAR processing for general users. Generally, the L0 data contain all the metadata necessary for processing, including the orbital state vectors and the radar parameters. The Level 1 (L1) products consist of the focused data. L1 products are the aforementioned SLC and GRD. Level 2 data consist of geophysical products that have been properly geolocated. After the SAR processing, the data must be geometrically and radiometrically calibrated, with the aim to retrieve the backscattering coefficient and to correct the distortions that intrinsically affect radar acquisitions when an area with complex topography is observed.

It is useful to explain some concepts that allow a better understanding of SAR data in general.

The different processing steps and their order will greatly depend on the type of information one wants to derive for data imagery. If the phase information is not necessary, i.e. GRD data are used, a classic workflow would be:

- *Orbit file refining*: The orbital information is necessary for geolocation and Doppler parameter estimation, which is related to the focusing capabilities. The orbit state vectors contain the three coordinates of position and velocity of the satellite centre of mass with reference to the used ellipsoid. The orbit state vectors provided in the metadata of the SAR products are generally not accurate and can be refined using orbit propagation techniques to generate precise orbit files which are available days to weeks after the generation of the product. As it defines the relationship between ground and image coordinates, this first step improves accuracy of later orbit-based calibration steps.

- *Thermal noise removal*: The thermal noise is the additive noise contribution that affects SAR images. Thermal noise can be corrected if the noise characterisation has been performed previously. The metadata of Sentinel-1 products contain dedicated LUTs (Look-Up Tables) that allow thermal noise removal using annotated thermal noise characterisation.

- *Radiometric calibration (to sigma0/beta0)*: Converts from digital number (DN) into the scattering coefficient – which is the conventional measure of the strength of radar signals from a distributed target, usually expressed in dB. It is a normalised dimensionless number, comparing the strength observed to that expected from an area of 1 m^2 [29, 30]. Sigma0 shows the radar reflectivity per unit area in ground range, whereas beta0 (or radar brightness) represents the radar reflectivity per unit area in slant range. Beta0 contains a geometric impact on the radar backscatter, e.g. bright pixels in foreshortened areas, and is useful to compare images of different SAR sensors. Sigma0 is the usual measure used to evaluate the reflectivity conditions of the surface, independent from geometric effects – and often used in scientific interpretation of SAR imagery since it allows relating the brightness in an image pixel to physical quantities [31].

- *Radiometric terrain flattening (optional)*: This step removes geometry-dependent radiometric distortions; it normalises measured backscatter with respect to terrain slope, removing most of the radiometric distortions from the data that are introduced by surface topography. It is not required if there is not much topography in the area of study.

- *Multilooking (optional)*: Multilooking processing divides the full synthetic aperture into sub-apertures (looks). The looks are summed incoherently to reduce the speckle noise. After the multilooking processing, the phase information is lost. Multilooking uses spatial averaging to reduce image speckle noise and converts it to ground range, producing an image with a standard pixel size. This operation results into a reduction of image resolution, as well.

- *Speckle filtering (optional)*: Speckle is a phenomenon that affects SAR images. It happens because each resolution cell associated with an extended target contains several scattering centres whose returns interfere positively or negatively, causing light or dark image brightness. Speckle appears in SAR images as granular noise. Speckle filtering increases image quality by reducing speckle. When this procedure is done at an early processing stage, speckle is not propagated in ongoing processes (i.e. terrain correction or conversion to dB). This is often regarded as an optional step, depending on the application. For example, speckle filtering is not advisable when there is the interest on identifying small spatial structures or image texture since it removes information [32]. Unlike multilooking, this step does not reduce spatial resolution.

- *Geocoding – range Doppler terrain correction*: The varying angles that SAR data is sensed result in images with some distortion related to side-looking geometry. Terrain correction is intended to compensate for these distortions so that the geometric representation of the image will be as close as possible to the real world. Range Doppler terrain correction is a correction of geometric distortions caused by topography, such as foreshortening and shadows, using a digital elevation model (DEM) to correct the location of each pixel [32].

2.6.3 Principles of optical sensors

The MSI on board of Sentinel-2, for example, measures the Earth's reflected radiance in 13 bands (443–2 190 nm) from the VNIR to the shortwave infrared (SWIR).

Radiance

In radiometry, radiance is the radiant flux emitted, reflected, transmitted or received by a given surface, per unit solid angle per unit projected area. Spectral radiance is the radiance of a surface per unit frequency or wavelength, depending on whether the spectrum is taken as a function of frequency or of wavelength. These are directional quantities.

Radiance is useful because it indicates how much of the power emitted, reflected, transmitted or received by a surface will be received by an optical system looking at that surface from a specified angle of view. For real, passive, optical systems, the output radiance is equal to the input, unless the index of refraction changes.

MSI has a wide swath (290 km) and its spatial resolution is dependent on the particular band. The three visible classical RGB bands and the VNIR bands have 10 m resolution, four bands in the VNIR vegetation and red edge and two wider SWIR bands have 20 m resolution and it is 60 m for three atmospheric correction bands.

2.6.4 Classic processing workflows for optical data

Sentinel-2 products can also be classified into three levels. Level 0 is compressed raw data, and it contains all the information required to generate the other product levels. This level, as well as Level 1A and Level 1B (uncompressed raw data with spectral bands coarsely coregistered and ancillary data appended, and radiometrically corrected radiance data, respectively) are not disseminated to users. The Level 1C are orthorectified top-of-atmosphere (TOA) reflectance with sub-pixel multispectral registration, and Level 2A provides orthorectified bottom-of-atmosphere reflectance, with sub-pixel multispectral registration. Level 1C and 2A products' images are divided into 100 km tiles, which can be fully or partially covered by image data. Partially covered tiles correspond to those at the edge of the swath [33].

- *Re-sampling*: It is common practice to re-sample all satellite bands to a unique pixel size (e.g. 10 m) to simplify the processing workflow.

- *Atmospheric corrections*: or a TOA correction. This processing is done in order to separate the actual reflectance that was emitted from the objects on the Earth's surface from the atmospheric disturbances that are part of the reflected energy recorded by the sensor. For the Level 2A products, this step is not necessary as atmospheric correction has already been applied.
- *Cloud masking*: Clouds present on the Sentinel-2 scenes can be mask out (using the quality scene classification band). Clouds are often irrelevant for ocean and land applications and this step enables cloudy and cloud-free pixels to be identified.
- *Indices*: From several bands one can derive numerous band ratios and indices (offering several advantages over stand-alone spectral bands), useful for a number of applications such as mapping purposes. Indices offer several advantages over stand-alone spectral bands (such as reduced effect of atmospheric conditions, canopy geometry and shading, decreased effect of soil background on canopy reflectance and enhanced variability of spectral reflectance).

2.6.5   Applications

Remote sensing has been applied to a wide range of applications in many fields such as agriculture, geology, meteorology, urban infrastructure, oceanography, archaeology and military. Different analysis techniques ranging from visual interpretation to the integrated use of remotely sensed data and other recent techniques of artificial intelligence have been used in recent years.

2.6.5.1   An application of SAR data: the interferometry

The most popular technique involving the use of SAR data is the interferometry, also known as InSAR (interferometric synthetic aperture radar). The InSAR is a radar technique that uses two or more SAR images to generate maps of surface deformation or DEM, using differences in the phase of the signals returning to the satellite. This technique can potentially measure millimetre-scale changes in deformation. It has applications for geophysical monitoring of natural hazards, e.g. earthquakes, volcanoes and landslides, and in structural engineering, in particular monitoring of subsidence and structural stability.

The interferometry uses differential phase of the reflected radiation, from multiple passes along the same trajectory. Since the source wave is produced by the satellite, the phase is known and can be compared to the phase of the return signal. The output of the interferometry is commonly called interferogram. The phase of the return wave depends on the distance to the ground, since the path length to the ground and back will consist of a number of whole wavelengths plus some fraction of a wavelength. This is observable as a phase difference or phase shift in the returning wave.

In practice, the phase of the return signal is affected by several factors, which together can make the absolute phase return in any SAR data collection essentially arbitrary, with no correlation from pixel to pixel. To get any useful information from the phase, some of these effects must be isolated and removed, by applying different filtering techniques.

The most important factor affecting the phase is the interaction with the ground surface. The phase of the wave may change on reflection, depending on the properties of the material. The reflected signal back from any one pixel is the summed contribution to the phase from many smaller 'targets' in that ground area, each with different dielectric properties and distances from the satellite, meaning the returned signal is arbitrary and completely uncorrelated with that from adjacent pixels.

Once the ground effects have been removed, the major signal present in the interferogram is a contribution from orbital effects. The interferometry expected that the satellites must be as close as possible to the same spatial position when the images are acquired. This means that images from two satellite platforms with different orbits cannot be compared. In practice, the perpendicular distance between them, known as the baseline, is often known to within a few centimetres but can only be controlled on a scale of tens to hundreds of metres. This difference causes a regular difference in phase that changes smoothly across the interferogram and can be modelled and removed. The difference in satellite position also alters the distortion caused by topography, meaning an extra phase difference is introduced by a stereoscopic effect. The longer the baseline, the smaller the topographic height needed to produce a fringe of phase change, known as the altitude of ambiguity. This effect can be exploited to calculate the topographic height and used to produce a DEM.

Once the ground, orbital and topographic contributions have been removed, the interferogram contains the deformation signal, along with any remaining noise. The signal measured in the interferogram represents the change in phase caused by an increase or decrease in distance from the ground pixel to the satellite, therefore only the component of the ground motion parallel to the satellite line of sight vector will cause a phase difference to be observed. One fringe of phase difference is generated by a ground motion of half the radar wavelength. Phase shifts are only resolvable relative to other points in the interferogram. Absolute deformation can be inferred by assuming one area in the interferogram (e.g. a point away from expected deformation sources) experienced no deformation, or by using a ground control to establish the absolute movement of a point.

Interested reader can refer to [34–41] for a detailed description of the topic.

2.6.5.2   An application of optical data: the object-based image analysis

The object-based image analysis (OBIA) technique has recently acquired an important role in many applications based on machine learning techniques [42]. For this reason we have decided to briefly introduce this technique as an example of application of optical data.

Urban feature (e.g. roads and buildings) extraction is used for planimetric base mapping, the generation of detailed topography, mapping land cover and land use, urban forestry assessment, infrastructure condition assessment, development monitoring, updating of old cartography and emergency response and disaster management. Urban environments are very difficult to map from computer processing due to the presence of many different environmental materials. In fact a map of land use is largely based on visual interpretation, which is subjective as well as time-consuming and costly. Fortunately, in recent years the development of new technologies has provided important tools to minimise visual interpretation and to obtain a better classification accuracy at a high level of detail.

Detailed and timely information of urban territory can be useful and is requested in many cases, such as after natural or human disasters, to provide a quick assessment to help urgent recovery actions. High spatial resolution imagery provides more detailed information and for this reason it is often preferred over moderate resolution. In the past decade, the highest spatial resolution of available data has increased and current satellite missions provide images with spatial resolutions in the range of 0.5–2.5 m.

As a result of greater spatial complexity and the growing availability of these data, the search for new object-oriented classification systems has been encouraged that take better advantage of spatial features. This approach has proven to be able to provide better classification results than traditional pixel-based classification approaches, especially for fine spatial resolution data and in heterogeneous landscapes such as urban environments [43]. Unlike the traditional pixel-based technique, OBIA encloses semantic information not in the individual pixel but in groups of pixels with similar characteristics, such as colour, texture, brightness and shape, by minimising the negative impact of high intraspectral variation.

Two stages are involved in the OBIA method: segmentation and classification. Objects are created through the segmentation step, which divides the image into homogeneous regions (objects), grouping pixels that have similar semantic or geometric characteristics. Classification is then implemented based on objects instead of individual pixels.

In addition to information related to each object (e.g. spectral signature, shape and size), there will also be information related to interrelationships among the objects. The specific characteristics that distinguish objects of interest (for instance, buildings) from others need to be identified for classification. An example of building extraction with the OBIA technique and very high-resolution satellite data for updating old topographic maps is presented in [44].

Moreover, in [45] the authors have combined the object-based classification method with deep learning to improve the interpretation accuracy of high-resolution images. As regards machine learning techniques, deep learning offers another efficient strategy to fill the gap between complex images and the need for timely and accurate classification. However, due to the hierarchical abstract nature of deep learning, it is difficult to capture the precise outline of different objects at the pixel level. Experimental results indicate that this combination is effective for mapping complex urban image datasets.

2.7 Conclusions

The principles underlying remote sensing and the techniques used for data processing were discussed in this chapter. The focus was then placed on the satellites of the Copernicus programme and specifically on two particular platforms, Sentinel-1, carrying on board an active SAR sensor, and Sentinel-2, carrying on a multispectral optical sensor. This choice has been made because, the case studies which will be presented later in this book will be based essentially on these two types of data.

References

[1]   What is remote sensing? [ESA Eduspace page]. Paris: European Space Agency. Available from www.esa.int/SPECIALS/Eduspace_EN/SEMF9R3Z2OF_0. html.

[2]   Chuvieco E. *Fundamentals of Satellite Remote Sensing: An Environmental Approach*. London: CRC press; 2016.

[3]   Sabins Jr F.F. *Remote Sensing–Principles and Interpretation*. Switzerland: WH Freeman and company; 1987.

[4]   Rees G., Rees W.G. *The Remote Sensing Data Book*. Cambridge: Cambridge university press; 1999.

[5]   Colwell R.N., Simonett D.S., Estes J.E. *Manual of remote sensing: v. 1: Theory, instruments and techniques.-v. 2: Interpretation and applications*. Falls Church, VA (USA): Amer. Soc. of Photogrammetry; 1983.

[6]   Elachi C., Van Zyl J.J. *Introduction to the Physics and Techniques of Remote Sensing*. **28**. John Wiley & Sons; 2006.

[7]   Born M., Wolf E. *Principles of Optics: Electromagnetic Theory of Propagation, Interference and Diffraction of Light*. Elsevier; 2013.

[8]   Shankar R. *Fundamentals of Physics II: Electromagnetism, Optics, and Quantum Mechanics*. Yale: Yale University Press; 2020.

[9]   Campbell J.B., Wynne R.H. *Introduction to Remote Sensing*. Guilford Press; 2011.

[10] European Space Agency. Radiometric resolution. Available from https://sentinel.esa.int/web/sentinel/user-guides/sentinel-2-msi/resolutions/radiometric.

[11] European Union's Earth Observation Programme. Copernicus Programme. 2020. Available from www.copernicus.eu/en.

[12] Di Martire D., Confuorto P., Frezza A., *et al*. 'X- and C-band SAR data to monitoring ground deformations and slow-moving landslides for the 2016 Manta and Portoviejo earthquake (Manabi, Ecuador)'. *2018 IEEE International Conference on Environmental Engineering (EE)*; 2018. pp. 1–6.

[13] Ullo S.L., Angelino C.V., Cicala L., *et al*. 'SAR interferometry with open Sentinel-1 data for environmental measurements: The case of Ischia earthquake'. *2018 IEEE International Conference on Environmental Engineering (EE)*; 2018. pp. 1–8.

[14] Ullo S.L., Langenkamp M.S., Oikarinen T.P., *et al.* 'Landslide geohazard assessment with convolutional neural networks using sentinel-2 imagery data'. *In: IGARSS 2019 - 2019 IEEE International Geoscience and Remote Sensing Symposium*; 2019. pp. 9646–9.

[15] De Corso T., Mignone L., Sebastianelli A. Application of DInSAR technique to high coherence satellite images for strategic infrastructure monitoring. IGARSS; Virtual Symposium, September 26, October 2, 2020; 2020.

[16] Ullo S.L., Addabbo P., Martire D.D., *et al.* 'Application of DInSAR technique to high coherence sentinel-1 images for dam monitoring and result validation through *in situ* measurements'. *IEEE Journal of Selected Topics in Applied Earth Observations and Remote Sensing*. 2019;**12**(3):875–90.

[17] Cicala L., Angelino C.V., Fiscante N., *et al.* Landsat-8 and sentinel-2 for fire monitoring at a local scale: a case study on Vesuvius. p. 1–6. 2018. Available from https://ieeexplore.ieee.org/document/8385269/.

[18] Addabbo P., di Bisceglie M., Focareta M. Combination of LANDSAT and EROS-B satellite images with GPS and LiDAR data for land monitoring. A case study: The Sant'Arcangelo Trimonte dump. IGARSS; Milan, Italy, 2015; 2015. pp. 882–5.

[19] Ullo S., Angelino C.V., Cicala L., *et al.* Use of differential interferometry on sentinel-1 images for the measurement of ground displacements. Ischia Earthquake and Comparison with INGV Data. p. 2216–2219. 2018. Available from https://ieeexplore.ieee.org/document/8518715/.

[20] Addabbo P., Focareta M., Marcuccio S., Votto C., Ullo S.L. 'Contribution of Sentinel-2 data for applications in vegetation monitoring'. *Acta Imeko*. 2016;**5**(2):44–54.

[21] Addabbo P., Focareta M., Marcuccio S. Land cover classification and monitoring through multisensor image and data combination. IGARSS; Beijing, China, July 10-15; 2016.

[22] Focareta M., Marcuccio S., Ullo S. Combination of Landsat 8 and sentinel 1 data for the characterization of a site of interest. A case study: the Royal Palace of Caserta. 1st international conference on metrology for archaeology.; Benevento, Italy; 2015.

[23] Luca C., Sara P., Cesario Vincenzo A., *et al.* Unsupervised post-fire assessment of burned areas with free and open multispectral data using OBIA. 2018. Available from www.geobia2018.com,Ωhttps://hal.archives-ouvertes.fr/hal-01957184.

[24] Fiscante N., Focareta M., Galdi C. Analysis and validation of high-resolution satellite DEMs generated from EROS-B data for Montaguto landslide. IGARSS; Munich, Germany, July 2012; 2012. pp. 7549–52.

[25] Beltramonte T., Di Bisceglie M., Galdi C., *et al.* 'CFAR detection of fire events in non-homogeneous non-Gaussian background'. *2012 Tyrrhenian Workshop on Advances in Radar and Remote Sensing*. 2012:252–6.

[26] Ulaby T., Moore K., Fung K. 'Microwave remote sensing'. *Microwave remote sensing fundamentals and radiometry*. 1981;**1**.

[27] Woodhouse I.H. *Introduction to Microwave Remote Sensing*. CRC Press; 2005.

[28] Cumming I., Wong F. *Digital Processing of SAR Data*. Norwood: Artech House; 2005.

[29] European Space Agency. ASAR Handbook. 2020. Available from https://earth.esa.int/handbooks/asar/CNTR5-3.html#eph.asar.gloss.prodt:SIGMA [Accessed August 2020].

[30] European Space Agency. Sentinel 1 SAR Definitions. 2020. Available from https://sentinel.esa.int/web/sentinel/user-guides/sentinel-1-sar/definitions.

[31] Airubs. Calculation of beta naught and sigma naught for TerraSAR-X data. 2008. Available from https://www.intelligence-airbusds.com/files/pmedia/public/r465_9_tsxx-airbusds-tn-0049-radiometric_calculations_d1.pdf.

[32] Filipponi F. 'Sentinel-1 GRD preprocessing workflow'. *Proceedings*. 2019;**18**(1):11.

[33] European Space Agency. Sentinel-2 User Handbook. 2015. Available from https://sentinel.esa.int/documents/247904/685211/Sentinel-2_User_Handbook.

[34] Ferretti A., Monti-Guarnieri A.V., Prati C., *et al. INSAR Principles B*. ESA Publications; 2007.

[35] Rucci A., Ferretti A., Monti Guarnieri A., Rocca F, Guarnieri A.M. 'Sentinel 1 SAR interferometry applications: the outlook for sub millimeter measurements'. *Remote Sensing of Environment*. 2012;**120**(4):156–63.

[36] Ferretti A., Prati C., Rocca F. 'Permanent scatterers in SAR interferometry'. *IEEE Transactions on Geoscience and Remote Sensing*. 2001;**39**(1):8–20.

[37] Rocca F., Prati C., Monti Guarnieri A., Ferretti A. 'SAR interferometry and its applications'. *Surveys in Geophysics*. 2000;**21**(2-3):159–76.

[38] Ferretti A., Prati C., Rocca F., *et al.* 'Multibaseline SAR interferometry for automatic DEM reconstruction (DEM)'. *Third ERS Symposium on Space at the service of our Environment*. 1997;**1809**:414.

[39] Rocca F. 'An overview of SAR interferometry'. *Proceedings of the 3re ERS Symposium*. **1997**. Florence; 1997.

[40] Prati C., Giani M., Leuratti N. 'SAR interferometry: a 2-D phase unwrapping technique based on phase and absolute values informations'. *10th Annual International Symposium on Geoscience and Remote Sensing*. IEEE; 1990. pp. 2043–6.

[41] Prati C., Ferretti A., Perissin D. 'Recent advances on surface ground deformation measurement by means of repeated space-borne SAR observations'. *Journal of Geodynamics*. 2010;**49**(3-4):161–70.

[42] Yang H. *Artificial Intelligence Science and Technology: Proceedings of the 2016 International Conference (AIST2016)*. USA: World Scientific Publishing Co., Inc; 2017.

[43] Weng Q. *Remote Sensing and GIS Integration: Theories, Methods, and Applications*. McGraw-Hill Education; 2009.

[44]   Ullo S.L., Zarro C., Wojtowicz K., Meoli G., Focareta M. 'LiDAR-based system and optical VHR data for building detection and mapping'. *Sensors.* 2020;**20**(5):1285.

[45]   Zhao W., Du S., Emery W.J. 'Object-based convolutional neural network for high-resolution imagery classification'. *IEEE Journal of Selected Topics in Applied Earth Observations and Remote Sensing.* 2017;**10**(10):3386–96.

Chapter 3

Artificial intelligence, machine learning and deep learning

Maria Pia Del Rosso[1], Silvia Liberata Ullo[1], Alessandro Sebastianelli[1], Dario Spiller[2], Erika Puglisi[3], Diego Di Martire[4], Sara Aparicio[5], and Pia Addabbo[6]

The aim of this chapter is to introduce the reader to the concepts of artificial intelligence (AI), a branch of computer science attempting to build machines capable of intelligent behaviour, as well as its subdisciplines, machine learning (ML) and deep learning (DL). By highlighting the differences between ML and DL and tracing the steps that led to their developments, this chapter explores the ability of a machine to learn instead of being explicitly programmed. This chapter focuses on AI and its related disciplines. Some applications based on DL and deep neural networks will be presented as main case studies later in the book.

3.1 Introduction

The concepts that laid the foundation for AI can be traced back to the 1600s, when the first examples of powerful calculators appeared on the scene: machines which were able to execute mathematical operations faster and in a more accurate way than humans. The real idea appeared three centuries later, when Alan Turing and John McCarthy introduced the term 'thinking machines', by which they designated 'machines that could think autonomously'. The birth of the expression 'artificial intelligence' occurred finally in 1956, during a conference organised to create a machine that could emulate human intelligence and learning. As AI systems evolved, machines based on AI are now referred to as 'machines that respond to stimulation

[1]Engineering Department, University of Sannio, Italy
[2]Joint ASI - Φ-Lab, Italian Space Agency (ASI) and European Space Agency (ESA), Italy
[3]Φ-Lab, European Space Agency (ESA), Italy
[4]University of Naples Federico II, Italy
[5]Solenix/ESA, Italy
[6]University Giustino Fortunato, Italy

consistent with traditional responses from humans, given the human capacity for contemplation, judgement and intention' [1]. As the word implies, machine intelligence is artificial, programmed by humans to perform human activities.

AI systems can be rearranged into different groups, by taking into consideration their capabilities and the types of tasks they can solve. While general AI systems can solve problems intelligently (e.g. AI-powered stock trading system), narrow AI systems can perform specific tasks very well (e.g. AI-powered manufacturing arms).

According to Darrell M. West's report for Brookings Institute, these systems have three qualities: intentionality, intelligence and adaptability [1]. These three features are defined as follows:

- *Intentionality*: Humans design AI systems with the intention of making decisions from either historical data, real-time data or both. Consequently, AI systems contain predetermined responses.
- *Intelligence*: AI systems are able to implement decisions, taking advantage of different algorithms applicable in these fields (e.g. ML and DL). While this type of decision-making is not comparable to that of a human, it is the best approximation of machine to human intelligence.
- *Adaptability*: AI systems have the ability to learn and adapt as they receive information and make decisions. As AI systems learn from real-time data, they can refine their decision-making capabilities to improve the outcome.

3.1.1   Historical pills on the birth of AI

The official date of birth of AI is 1956, the year of the famous summer seminar held at Dartmouth College of Hanover (New Hampshire). At this occasion, the new discipline was founded and the term 'artificial intelligence' coined. This conventional date of birth, unanimously agreed upon the scientific community, might suggest that the question of AI origins can be definitively resolved. However, when talking about the history of AI, it is necessary to refer to cybernetics and to the advent of the first electronic calculators [2].

In 1623, Willhelm Sickhart developed a machine capable of performing mathematical calculations using numbers up to six digits, even if not independently. In 1642, Blaise Pascal built a machine capable of doing automatic reporting during calculations; and in 1674, Gottfried Wilhelm von Leibneiz created a machine capable of calculating sums, differences and multiplications in a recursive way. Nevertheless, it was only several years later, in 1837, when Charles Babbage described a historical novelty: the 'analytical engine' [3]. This term encompassed the first prototype of a mechanical computer, developed to perform generic tasks. The machine was never realised for economical reasons, but this project was undoubtedly the precursor of the modern personal computer.

Another century had to pass for the next important steps. In 1936, Alan Turing defined the concepts of 'computability' and 'Turing machine' [4]; and in 1943, Warren McCulloch and Walter Pitts created the first model of artificial neurons [5].

Figure 3.1   Participants of the 1956 Dartmouth Summer Research Project on AI in front of Dartmouth Hall. In the photo there are also Claude Shannon (front right), John McCarthy (back right), Marvin Minsky (back centre) and Nathaniel Rochester (back left). Photo credit Margaret Minsky via www.achievement.org.

Finally, in the famous summer 1956, a group of scientists, including John McCarthy (assistant professor in Dartmouth), Marvin Minsky (junior fellow in mathematics and neurology at Harvard), Nathaniel Rochester (research director in IBM) and Claude Shannon (mathematician in the Bell Telephone Laboratories), laid the foundations of what we call today 'artificial intelligence' (Figure 3.1).

The following years were charged with high expectations: Allen Newell, Bernard Shaw and Herbert Simon improved the *Logic Theorist*, a computer program able to prove theorems starting from mathematical principles, by creating the *General Problem Solver*, which was able to duplicate human problem-solving processes. In 1958, John McCarthy described an ideal program, the *Advice Taker*, able to find solutions for different, also non-mathematical, kinds of problems [6]. In 1959, Arthur Samuel created an algorithm able to play checkers and managed to improve its performance from past gaming experiences. This can be considered to be the birth of *ML*. However, all these projects and ideas started to encounter some difficulties: not only did adequate methods for simple cases turn out to be useless for more complex contests, but promising theories ended up not being applicable in practice, due to the low computational power and hardware limitations of the machines available at that time. Moreover, McCarthy's document [6] highlighted the negative aspects and limitations of artificial neurons, thus causing ML and neural networks to fall into disuse.

Scientists then had to reduce their expectations and focus on approaches more practically related to the work environment where computers and algorithms needed to be applied. This is why in the 1980s AI started to appear in factories. The first one was R1, used by *Digital Equipment* in 1982, which was able

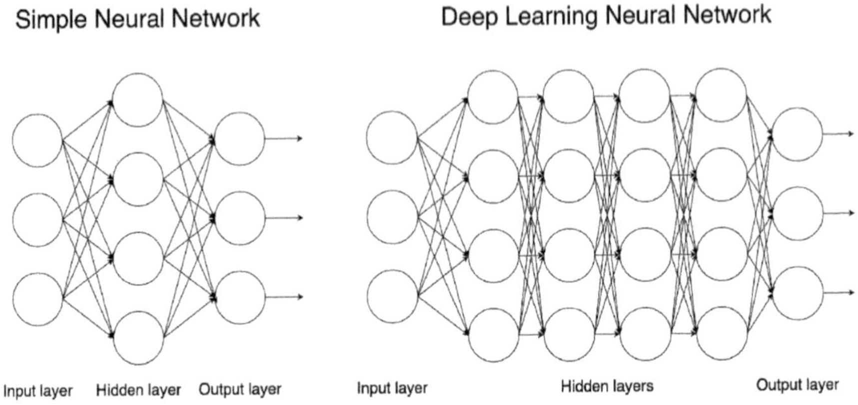

Simple Neural Network   Deep Learning Neural Network

Input layer   Hidden layer   Output layer   Input layer   Hidden layers   Output layer

Figure 3.2   Difference between a simple neural network and a deep neural
network

to determine if a customer's order needed to be modified for reasons of system functionality. Since the use of R1 made the company save a million dollars, soon many other factories followed its example. During the same period, in 1985, a group of researches reinvented (and successfully applied) the 'back-propagation' algorithm. This learning method for neural networks had been created in 1969 by Bryson and Ho, but since it had been not applicable at that time it had been abandoned.

In the 1990s, ML was rediscovered as a separate discipline from AI. Its purpose had changed from the development of AI for data-based, solution-oriented research for practical problems, influenced by statistical methods and probability calculus. Neural networks, a graph and neurons-based collection of advanced algorithms employed in ML for data modelling by using graphs of neurons started to be used in several fields and for different kinds of problems. This created the need of a more complex structure that was able to process many-faceted problems: the deep neural network.

In Figure 3.2 a simple neural network and a deep neural network are shown, where the difference is in the number of the hidden layers. More hidden layers lead to more connections between the inner units and thus to a structure more similar to the human brain. Later in the book we will talk in depth about how deep neural networks work and how they are configured. This kind of neural networks is able to recognise and distinguish objects and texts in images and videos. In 2006, Geoffrey Hinton was the first person who used the term *deep learning* to describe these networks.

Nowadays, AI is part of everyday life and even if it is not even remotely comparable to human intelligence, it is able to perform specific tasks in specific environments with optimal results.

3.2   Artificial intelligence, machine learning and deep learning

With reference to Figure 3.3, AI relates to any technique enabling computers to mimic human behaviour. Within the AI domain, ML and DL are defined as specialised applications. The next paragraphs intend to give more detailed information with this regard.

3.2.1   Machine learning

ML is an application of AI, providing AI systems with the ability to automatically learn from the environment and apply decision-making paradigms. There are a variety of algorithms that ML uses to learn iteratively, as well as describe and improve data in order to predict better outcomes. These algorithms use statistical techniques to spot patterns and then perform actions on these patterns. Even though neural networks and DL models are increasingly attracting the attention of researches and companies involved with AI, ML algorithms are still commonly used today as they show a good performance, especially with reduced dimension datasets. Moreover, ML algorithms have become widely diffused. This has been thanks to the growing availability of libraries in programming languages such as Python (e.g. *Scikit-learn*) or R (e.g. *Caret*), which provide rapid and efficient implementations of classification and regression models. It is noteworthy that ML algorithms usually require a preliminary feature engineering on the input data in order to extract and prepare the features for the model. This book will not cover ML algorithms in detail, but the interested reader may refer to several books in the literature, e.g. [7–9].

3.2.2   Deep learning

DL is a subset of ML techniques and represents the newest generation of AI. DL models can make their own predictions entirely independent of humans as they do not require any preliminary feature engineering (apart from data normalisation). These models are based on artificial neural networks, and *deep* refers to the high number of hidden layers included in the network. The design of these models is inspired by the biological neural network of the human brain involving complicated connections among billions of neurons. Indeed, a DL model infers conclusions from

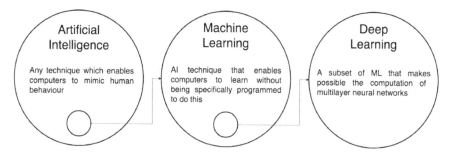

Figure 3.3   Differences and relationships between AI, ML and DL

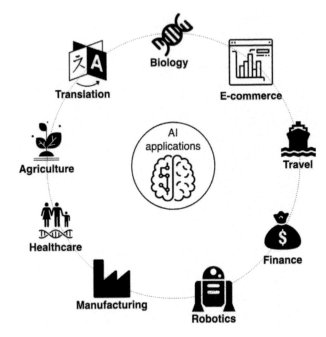

Figure 3.4   Some of the possible AI applications

input data trying to imitate what humans can do. As mentioned before, an in-depth discussion of neural networks will follow in the next chapter.

Just to anticipate some insights, it is worth highlighting that a neural network uses multiple layers to progressively extract (or understand) higher-level features from raw data. Taking image recognition applications as an example, DL algorithms will go beyond just recognising matrix pixels. Instead, they will detect edges at a certain level, noses at another and faces still at another level. With the ability to understand data from the lower level all the way up the chain, a DL algorithm can improve its performance over time and arrive at a decision.

The power of DL algorithms lays in their ability to take on both supervised and unsupervised learning tasks. They also approximate many development theories of the human brain. Nowadays, they are used by systems in areas as diverse as computer vision, speech recognition, natural language processing, audio recognition, bio-informatics and medical image analysis [10–14].

3.2.3   Real-life applications

AI is better suited for solving problems in certain areas than in others (see Figure 3.4). As in real life, problems are rarely simple, the most efficient use of AI is often limited to certain steps in solving a problem, while the rest is left to humans. For instance, AI-enabled chatbots might be able to follow employees up on their projects, to receive updates on their status, but managers still have to build, inspire and steer the teams in the right directions. A list of possible problems best suited for AI to solve are the following:

- *Repetitive tasks*: Manual tasks that follow logical steps to lead to a conclusion (e.g. packaging goods to be ready for delivery at a warehouse).
- *Data-intensive tasks*: Tasks involving the analysis of large amounts of data looking for patterns and anomalies (e.g. detection of fraud from financial records).
- *Superhuman tasks*: Tasks that require superhuman capabilities and speak to the limitations of human sensory skills and fine motor skills (e.g. robot surgeon can use the most precise movements to perform non-invasive surgeries; finely tuned computer vision can spot tumours on an MRI (magnetic resonance imaging) when human eyes cannot see it).

As AI systems advance, new challenges arise by making humans confront themselves with their limitations. While AI brings more efficiency into human lives, new problems are faced, created by integrating AI into everyday life. Only with more understanding and less fear, we can empower ourselves to move forward to the age of AI.

3.3   Supervised, unsupervised and reinforcement learning

As stated earlier, an ML algorithm is an algorithm that is able to learn from data. This circumstance enables the algorithm to solve tasks that are too difficult to solve with fixed programs written and designed by humans. Nevertheless, in order to evaluate the abilities of an ML algorithm, a quantitative measure of its performance must be designed. In this respect, a more formal description and some definitions are needed.

Definition
A computer program is said to learn from experience E with respect to a certain class of tasks T and performance measure P, if its performance at that class of tasks T, as measured by P, improves with experience E [15]. Usually P is specific to the class of tasks T being carried out by the system.

In a nutshell, it is worth highlighting that ML algorithms make decisions based on what they can learn from the environment. Therefore, this operation can be performed differently depending on the characteristics of the learning process [1]. Based on the way the learning process in an ML algorithm is devised, they can be categorised as supervised, unsupervised and reinforcement learning (RL) [13].

3.3.1   Supervised learning

Supervised learning is characterised by a function $f(x)$ that maps the input data x to the desired output data Y.

$$f(x) \rightarrow Y$$

(3.1)

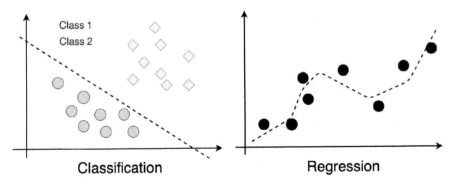

Figure 3.5   *Typical examples of algorithms for supervised learning*

It is called 'supervised' because the values of x and Y are known (i.e. the data are labelled) and, during the algorithm learning process, the predictions done on the training data are corrected using the Y values. The objective of a supervised learning algorithm is to find the mapping function f that achieves acceptable levels of performance based on the user needs. The greatest concern is the generalisation ability of the mapping function f with regard to new data, where the performances are usually lower than the ones of the training dataset.

The most common types of supervised problems, classification and regression, differ on the type of output variables. If the output variable is categorical, then it is a classification problem (e.g. colour can be red, blue, purple, etc.). If the output variable is a real value, then it is a regression problem (e.g., height can be on a scale of 0–10 m). Typical examples of algorithms for supervised learning are (see Figure 3.5):

- Classification, where the objective is to find the boundaries separating different categories. On the left of Figure 3.5, an example based on data with two features is shown.
- Regression, where the objective is to find the function mapping a set of input values to a set of output values. On the right of Figure 3.5, an example mapping a one-dimensional input to a one-dimensional output is shown.

In order to solve a given problem of supervised learning, one has to perform the following steps:

1. **Determine the type of training examples.** Before doing anything else, the user should decide what kind of data is to be used as a training set. For remote sensing applications, you usually deal with images or tabular data.
2. **Gather a training set.** The training set needs to be representative of the real-world use of the function. Thus, a set of input objects is gathered together with a corresponding set of outputs, either from human experts or from measurements.

3. ***Determine the input feature representation of the learned function.*** The accuracy of the learned function depends strongly on how the input object is represented. Typically, the input object is transformed into a feature vector, which contains a number of features that are descriptive of the object. The number of features should not be too large, because this would result in a curse of dimensionality, but it should contain enough information to accurately predict the output.

4. ***Determine the structure of the learned function and corresponding learning algorithm.*** For example, the user may choose to use support vector machines or decision trees.

5. ***Complete the design. Run the learning algorithm on the gathered training set.*** Some supervised learning algorithms require the user to determine certain control parameters. These parameters may be adjusted by optimising performance on a subset (called a validation set) of the training set, or via cross-validation.

6. ***Evaluate the accuracy of the learned function.*** After parameter adjustment and learning, the performance of the resulting function should be measured on a test set that is separate from the training set.

3.3.2 Unsupervised learning

Unsupervised learning, or self-organisation, is a type of ML that looks for previously undetected patterns in a dataset with no pre-existing labels and with a minimum of human supervision. In contrast to supervised learning that usually makes use of human labelled data, unsupervised learning allows for modelling of probability densities over inputs [16]. Unsupervised learning does not require a correct set of outputs Y, i.e. there are no reference outputs. The objective here is to present the most interesting structure that best describes the input data.

There are two types of unsupervised problems, clustering and association. Clustering problems deal with discovering relationships or similarities inside the input data (e.g. grouping voting behaviours by gender), while association problems focus on discovering rules inside the input data (e.g. female voters tend to vote for female candidates). Typical unsupervised learning algorithms are (see Figure 3.6):

- ***Principal component analysis (PCA):*** Given a collection of points in two-, three-, or higher-dimensional space, a 'best fitting' line can be defined as one that minimises the average squared distance from a point to the line. The next best fitting line can be similarly chosen from directions perpendicular to the first. Repeating this process yields an orthogonal basis in which different individual dimensions of the data are uncorrelated. These basis vectors are called principal components, and related procedures principal component analysis. PCA is the simplest analysis among the true eigenvector-based multivariate ones. Often, its operation can be thought of in such a way as to reveal the internal structure of the data and the variance in the data themselves. If a multivariate dataset is visualised as a set of coordinates in a high-dimensional data space (one axis per

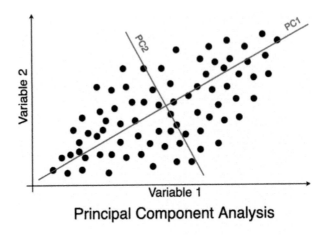

Principal Component Analysis

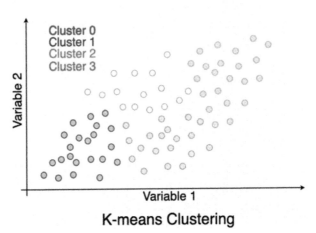

K-means Clustering

Figure 3.6   Typical examples of unsupervised learning algorithms

variable), PCA can supply the user with a lower-dimensional picture, i.e. a projection of this object when viewed from its most informative viewpoint. This is done by using only the first few principal components so that the dimensionality of the transformed data is reduced [17].

- *k-means clustering*: It is a method of vector quantisation, originally from signal processing, that aims to partition *n* observations into *k* clusters in which each observation belongs to the cluster with the nearest mean (cluster centres or cluster centroid), serving as a prototype of the cluster. This results in a partitioning of the data space into Voronoi cells. It is popular for cluster analysis in data mining [18].

3.3.3 Reinforcement learning

RL is concerned with how a software agent can find the best actions to take in an environment in order to maximise the notion of cumulative reward. The agent passes from one state (S) to another (S′) by choosing actions (A), these actions produce positive or negative changes in the environment, which are reported to the agent as rewards (R). The way in which the agent chooses the action is sequential and modelled as a map called policy (see (3.2)).

$$\pi : A \times S \to [0, 1] \tag{3.2}$$

The policy (π) represents the probability that the agent takes the action A in the state S. The expected, produced by the action chosen, is represented by the value that is influenced by the policy (see (3.3)).

$$V_\pi(S) = E[R] \tag{3.3}$$

The value function estimates how good it is to be in a particular state and the variable R in its formula represents the sum of all positive and negative rewards obtained along the path (see (3.4)).

$$R = \sum_{t=0}^{\infty} \gamma^t r_t \tag{3.4}$$

In (3.4), r_t is the reward obtained at step t and γ^t is the discount rate, a variable $\in [0,1]$, which denotes how much the reward at step t weighs in the cumulative reward; the further in the past the step, the lower the weight of its reward.

The final goal is to find a policy with the maximum expected return and a good balance between the exploration and exploitation phase. The exploration phase is similar to a 'test phase', in which the algorithm explores the environment by randomly choosing the actions and collecting the results of its choices. In the exploitation phase, the algorithm chooses the action with the best long-term effect, basing its choice on past learned experiences.

Typical RL algorithms are:

- **Q-learning:** This algorithm has a function, called Q-function, which calculates the quality of a state–action combination (the Q-function name does not refer to the normal distribution). Before learning begins, the function is initialised to an arbitrary value. At each time t, the agent selects an action a_t, observes the reward r_t, enters in a new state s_{t+1} (that depends on both the previous state s_t and the selected action), and the Q-function is updated by (3.5).

$$Q^{new}(s_t, a_t) \leftarrow Q(s_t, a_t) + \alpha(r_t + \gamma(\max_a Q(s_{t+1}, a)) - Q(s_t, a_t)) \tag{3.5}$$

Thus, it is updated using the old value of the Q-function plus the new weighted value multiplied for a learning rate α.

- **State–action–reward–state–action (SARSA):** the name of the algorithm reflects the variables used in the update rule, which are the current state (s), the action chosen (a), the reward obtained (r), the state in which the agent enters after taking that action (s') and the next action the agent chooses in its new state ($\acute{a}$). See (3.6).

$$Q^{\text{new}}(s_t, a_t) \leftarrow Q(s_t, a_t) + \alpha(r_t + \gamma(Q(s_{t+1}, a_{t+1})) - Q(s_t, a_t)) \qquad (3.6)$$

Unlike Q-learning, this algorithm also takes into account the next action the agent is going to choose in its new state. Q-learning can be described as an *on-policy* method because the Q-function is learned from actions chosen exclusively using the current policy. On the other hand, Q-learning algorithm is an *off-policy* method since the Q-function is learned choosing the action that maximise Q (this also leads to a random choice of actions) and consequentially no policy is needed.

3.4 Training and validation of a model

In ML, after designing and training phases, the user wants to evaluate the performances of the model. This operation requires the definition of a training and a test dataset. The former is used during the learning phase, whereas the latter is necessary for the evaluation of the performances on an independent dataset. Usually, the test dataset is from 20% up to 40% of the input dataset. This approach is represented on the top of Figure 3.7.

A more sophisticated splitting of the input dataset is the one based on training, validation and test dataset. In this case, training and validation datasets are used for training and optimisation of the hyper-parameters (e.g. number of trees in a random forest, number of layers in a neural network). After the definition of the model, the test dataset is used as an independent dataset for the final evaluation of the model performances. Usually, both the validation and the test dataset represent the 20% or the original input dataset. The training, validation and test splitting are represented on the bottom of Figure 3.7.

3.5 Cross-validation for supervised learning

The assessment of the performance accuracy when performing supervised learning can be a tricky task. Sometimes, the simple splitting of the dataset into training, validation and test is not enough to get a reliable estimation of the model performance. Indeed, in general, the bias and variance of the model can depend on the way the input dataset has been used for the training and validation phases.

To clarify this concept, consider a model that has been trained and optimised with respect to its hyper-parameters over a given classification problem. If training and hyper-parameter setting have been accomplished considering a fixed validation dataset and the final results are computed over an independent test dataset, the

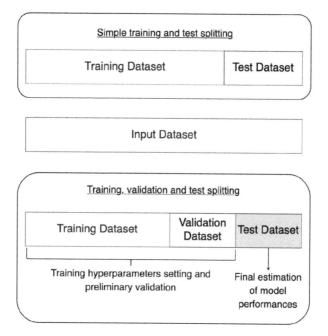

Figure 3.7   Definition of training, validation and test datasets

performance assessment can be misleading. Indeed, the user will tend to optimise the model considering the fixed splitting of the input dataset, thus risking to overfit the model with respect to the considered test dataset. Consequently, the performance of the model can be quite different on other unseen data.

To reduce the risk of overestimating the performance of the classification model, the performance assessment can be conducted by using the cross-validation scheme. One well-known technique is the k-fold cross-validation, depicted in Figure 3.8 in the case of $k = 10$. In this case, the evaluation of the model performances is carried out multiple times (10 times in the reported example) on different test sets. The final performance evaluation if provided as the mean value of each single perfor-mance. The suggested number of folds in literature is usually 5 or 10, but this can be a problem-dependent parameter, especially affected by the dimension of the input dataset.

It should be noted that most complicated approaches can be used as, for instance, the nested k-fold cross-validation, where the validation is mixed with the hyper-parameter selection. More details can be found in [19].

3.6   Class imbalance and data augmentation

Class imbalance and data augmentation are two important concepts that must be known by researchers and technicians working with ML. Even though they are sepa-rate concepts, data augmentation can help solving the problems introduced by the

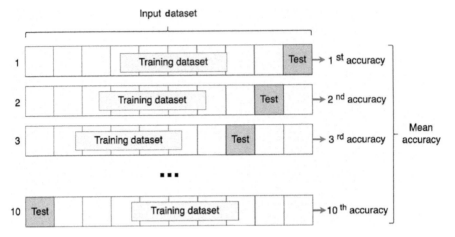

Figure 3.8   *Example of ten-fold cross-validation*

class imbalance. In this section, class imbalance and data argumentation will be first described separately, while the final part will deal with how to create balanced datasets using data augmentation. In the following section, supervised learning is considered, even though class imbalance can affect also unsupervised learning.

3.6.1   Class imbalance

When dealing with classification problems, most of the practical problems that a data scientist is asked to solve can be affected by class imbalance. In such cases, the number of samples for each class is not constant, and one or more classes can be dominating with respect to the others. Let a binary classification problem be considered. For instance, a dataset containing the number of healthy and sick patient with regard to a very rare disease is given for an entire country. In such case, it can happen that the number of healthy people represents the 99% of the dataset, with the remaining 1% containing the samples of sick people. Several strategies and differ-ent models can be developed to deal with such an imbalanced dataset, however let the following situation being considered. A model is trained that always predicts the input as a healthy individual. Given the distribution of the data, the accuracy of the model would be 99%, which can be interpreted as a very good result at first glance. However, this model is completely unable to detect sick people, hence is not useful at all as a medical tool.

The class imbalance problem can be tackled in different ways and it is still an open research field [20]. On the one hand, it has been shown that some ML models can be more robust than the other with regard to this issue, as in [21–24]. Usually, the class imbalance problem can be solved in one of the following ways:

- Set class weight/importance. In this case, the minority classes are given some higher weights than the majority class in the evaluation of the performance metrics.

- Use data augmentation. In this case, the data scientist tries to solve the problem by improving the distribution in the input dataset.
- Develop a specific algorithm to take into account the distribution of a specific dataset.

In all these cases, it is strongly suggested to perform a cross-validation campaign to properly predict the performances of the model.

3.6.2   Data augmentation

The best and easiest way to make a model generalise is to train it on a large amount of data. However, most of the time, for practical applications, we are provided with limited amounts of data. One way to counteract this issue is to create new, fake data and add them to our training dataset. For some domains, like classification problems, this is fairly straightforward and easy. A classifier needs to take a complicated, high-dimensional input x (usually an n-dimensional vector called tensor) and summarise it with a single category identity y. One of the main tasks a classifier is asked to face is to be invariant to a wide variety of transformations. We can easily generate new (x, y) pairs by transforming the x inputs in our training set according to different techniques, depending on the features and characteristics of the input dataset.

If you have a binary classification problem with 100 data rows in one class and 10 data rows in the other class, you could simply duplicate examples from the minority class in the training dataset prior to fitting a model. This can balance the distribution of the class but does not provide the model with any extra details. Instead, we use data augmentation, which can be very powerful. The synthesis of new examples from the minority class is an advancement over the replication of examples from the minority class.

3.6.2.1   For tabular data

SMOTE and ADASYN are popular techniques to generate new data and balance out the dataset classes.

SMOTE is short for *Synthetic Minority Oversampling Technique* and was introduced by Nitesh Chawla *et al.* in their 2002 paper named for the technique titled 'SMOTE: Synthetic Minority Over-sampling Technique' [25]. SMOTE works by selecting examples that are close in the feature space, drawing a line between them and selecting a new sample at a point along that line.

Another approach involves generating synthetic samples inversely proportional to the density of the examples in the minority class. That is, generating more synthetic examples in regions of the feature space where the density of minority examples is low, and fewer or none where the density is high. This modification to SMOTE is referred to as the *Adaptive Synthetic Sampling Method*, or ADASYN, and was proposed by Haibo He *et al.* in their 2008 paper named 'ADASYN: Adaptive Synthetic Sampling Approach for Imbalanced Learning' [26].

Note that Python offers a dedicated package (*imbalanced-learn*) to deal with imbalanced dataset, where SMOTE and ADASYN are implemented along with other approaches.

3.6.2.2   For image data

Dataset augmentation is a very popular approach for computer vision tasks such as image classification or object recognition as images are high dimensional and include an enormous variety of factors of variation, many of which can be easily simulated. Operations like translating the training images, a few pixels in each direction, rotating or scaling the image, can often greatly improve the generalisation ability. As of today, more advanced techniques are used in computer vision, such as the generative adversarial networks, to generate artificial training data for ML tasks.

3.6.3   Noise robustness

Noise is often introduced to the inputs as a dataset augmentation strategy. The addition of noise with infinitesimal variance at the input of the model is equivalent to imposing a penalty on the norm of the weights. Noise injection is much more powerful than simply shrinking the parameters, especially when the noise is added to the hidden units. Another way that noise has been used in the service of regularising models is by adding it to the weights. This technique has been used primarily in the context of recurrent neural networks. This can be interpreted as a stochastic implementation of Bayesian inference over the weights.

3.6.4   Choice and use of a dataset

The efficiency and correctness of a model strongly depends on the data with which it was trained. For this reason it is of vital importance to check that the dataset respects some characteristics:

- *Heterogeneity of the data.* If the feature vectors include features of many different kinds (discrete, discrete ordered, counts, continuous values), some algorithms are easier to apply than others. Many algorithms, including support vector machines, linear regression, logistic regression, neural networks and nearest neighbour methods, require the input features to be numerical and scaled to similar ranges (e.g. to the $[-1,1]$ interval).
- *Redundancy in the data.* If the input features contain redundant information (e.g. highly correlated features), some learning algorithms (e.g. linear regression, logistic regression and distance-based methods) will perform poorly because of numerical instabilities. These problems can often be solved by imposing some form of regularisation.
- *Presence of interactions and non-linearities.* If each of the features makes an independent contribution to the output, then algorithms based on linear functions (e.g. linear regression, logistic regression, support vector machines, naive Bayes) and distance functions (e.g. nearest neighbour methods, support vector machines with Gaussian kernels) generally perform well. However, if there are complex interactions among features, then algorithms such as decision trees and neural networks work better, because they are specifically designed to discover these interactions.

Another important aspect is the amount of data, as the larger the dataset, especially in remote sensing, the more the model will be able to learn all the possible aspects of a specific phenomenon. Obviously, the dataset must respect the rules listed above.

Chapter 6 will discuss more on datasets. In particular, techniques for the automatic or manual creation of datasets will be introduced, some common data processing techniques will be explained and some rules that must be respected when creating a dataset will be presented.

3.7 Applications

As previously explained there are various applications and different sectors in which the several AI techniques can be applied. The idea of this book, however, is to introduce an inexperienced user to the world of AI applied to remote sensing. Consequently, some applications of AI for EO (Earth observation) will be listed in the following subsections, and other applications will be presented in the following chapters as case studies. However, to give an idea of other possible applications, some AI applications will be listed in Section 3.7.2.

3.7.1 Remote sensing applications

AI approaches have been increasingly used to extract patterns and insights from the ever-increasing stream of geospatial data [27]. AI techniques can be applied to remote sensing data for all the applications that are common in that discipline, from weather forecasting and climate change parameters measurement to natural disasters monitoring. AI applied to remote sensing can also be used for improving social and economical aspects, such as urban development monitoring, cultural heritage preservation, air and maritime traffic management. In the next chapters, the crop-type mapping problem will serve as an example to illustrate how AI for EO can improve the maintenance of food delivery in poor nations, e.g. central Africa.

Since one of the most important aspects of remote sensing can be attributed to risk prevention, it is important to talk about how AI techniques can contribute to improve risk management. Obviously, we are talking of risks induced mainly by natural disasters.

3.7.1.1 Natural disasters

The management of the hazard and risk of natural disasters is based on different types of data, sources of information and types of models. These analyses involve large amounts of data from a wide range of sources. Experts may also find it difficult to develop models that allow them to understand the potential impacts of a hazard on the environment and the population. According to information provided by World Health Organization from 1900 to 2019, there are around 15 million examples for all types of disaster [28]. All natural disasters include those from drought, floods, extreme weather, extreme temperature, landslides, dry mass movements, wildfires, volcanic activity and earthquakes. In this context, the use of new methodologies and machines provides new ways of examining these complex relationships. It can potentially provide precise, efficient and useful answers in a pace which allows analysis and calculations that the human mind is not able to accomplish. Natural disasters are occurrences that are hard to prevent because they result from natural processes that cannot be forecast [29]. Such disasters affect millions of people every year and cause losses of human lives and property damage worth hundreds of billions. The impact of these phenomena could be reduced if we were able to predict their occurrence. There are different ways of decreasing the risks of natural disasters. One of them is implementing disaster reduction strategies. There are already several developed countries that apply the concept of 'disaster reduction'. In addition to disaster reduction programmes, there are several ways of forecasting or reducing the risks using AI technology. Some of them are Big Data, ML and DL. Utilising these methods facilitates tasks in visualising, analysing and predicting natural disasters [30, 31].

AI and ML have increasingly obtained recognition in recent years. As these techniques are proliferating in all fields of competence, hazard forecasting is no exception. New applications are being generated continuously, by not only improving the different components of risk modelling such as exposure, vulnerability and hazard, but also optimising the prioritisation of resources during damage response and reconstruction [32]. The first applications have tried to better understand the exposure to disasters from the physical (detecting seismic vulnerability in urban areas) as well as from the socio-economic side (social exposure and vulnerability). These kinds of applications have based principally on the analysis of satellite imagery features (using radar data and LiDAR). Newer applications are also starting to take advantage of artificial vision approaches to identify vulnerabilities from satellite imagery [33].

In the field of natural phenomena, ML has been used in many applications such as earthquakes, landslide susceptibility, identification of landslides, wildfires, groundwater flooding estimation and weather forecasting. Many of these applications use temporal data from hundreds of thousands of sensors (whether physical, like weather stations or earthquake stations, or remote, such as satellites) [34] and other geophysical characteristics to forecast hazard outcomes. A different approach involves examining the impact of hazards by carrying out what is called a hazard assessment. This procedure is implemented by training the prediction algorithm using past events as training data [35, 36]. Subsequently, once the models are trained, these algorithms are able to predict which areas are most likely involved in the event in question (landslide susceptibility, groundwater flooding susceptibility) [37, 38]. In recent years, the mapping of post-disaster events and the assessment of the damage associated with them are also emerging as new applications. Although it is not easy to use optical data from satellites, some approaches use very high-resolution optical images from unmanned aerial vehicles. In other cases, machines can easily make use of complex data which are difficult for humans to interpret, such as radar data (flooded extension mapping, landslide mapping) [39]. Other applications that have been very successful in recent years are related to land monitoring using satellite data. In fact, in many cases such as movements of slow landslides, control of the stability of buildings, bridges or dams, time series of satellite data are used to estimate the slow movements and possible deformations associated with the phenomenon [40]. In addition, in many cases the coupling of satellite data with ML algorithms allows rapid identification of certain events. In fact, some ML algorithms are already developed in a programming language and available in user interfaces, such as the image classification algorithms for remote sensing [41]. Moreover, there are several readily available ML algorithms inside remote sensing and GIS software packages, some of which are free such as Saga Gis or QGIS.

3.7.2 Other applications

While the application of ML in remote sensing is relatively new, other sectors of engineering have been using ML for years.

3.7.2.1 Self-driving cars

One of the most famous and advanced applications of ML is self-driving cars. Thanks to ML, these autonomous cars are very much capable of sensing the environment around them and moving safely, almost without the need for human intervention. They integrate a set of sensors whose measurements are used for extrapolating information, which feed ML models. Examples of tasks performed by AI are object detection, object identification/recognition, object localisation and movement prediction. The most common algorithms include regression, pattern recognition, clustering and decision matrices. Tasks related to data coming from a camera are part of the discipline that fall inside the so-called computer vision research field.

3.7.2.2 E-commerce and social media

Other applications that deserve to be mentioned are the use of ML algorithms in e-commerce sites and social media. E-commerce sites use them to compile a list of items for sale that may be interesting for the user, based on recent or most common purchases. On the other hand, social media use ML for face recognition in pictures or for suggesting posts to the users based on previous liked posts or followed accounts.

3.7.2.3 Space exploration

ML and DL have found very interesting applications in space exploration. For instance, crater detection in celestial bodies of the solar system has yielded valuable results. Further promising examples are reported considering Mars and the Moon [42–44], and this application can be extended as well to other celestial bodies such as asteroids. The practical implications of these studies are quite relevant with regard to the guidance, navigation and control systems of autonomous space probes for explorations.

3.7.2.4 Fault detection of navigation systems

Unsupervised ML techniques can be used to detect faults in navigation systems of complex systems such as orbiting spacecrafts [45–49]. This application can help to understand the reason behind faults by performing correlation analyses and predicting next faults based on time series analyses.

3.7.2.5   Medical applications

ML is increasingly used for medical applications [50]. For instance, ML and DL studies are reported for Alzheimer's disease [51], cerebrovascular disease [52] and brain tumour segmentation [53]. Very recent studies reportedly help to understand the spread of Covid19 [53–55].

3.8   Conclusions

In conclusion, topics based on AI have been addressed in this chapter. The fundamental notions involving AI and the resulting disciplines have been introduced, with a greater focus on ML. Furthermore, the use of these techniques, the fields of application and their great potential have been highlighted.

The next chapters will focus more on DL and its applications. This choice is due to the fact that most of the new applications to remote sensing are based on DL techniques, and in particular on convolutional neural networks , which will be discussed ahead in detail.

References

[1]   Wu J. *AI, machine learning, deep learning explained simply* [Article]. Towards data science. 2019. Available from https://towardsdatascience. com/ai-machine-learning-deep-learning-explained-simply-7b553da5b960 [Accessed August 2020].

[2]   Somalvico M., Amigoni F., Schiaffonati V. *Intelligenza artificiale*. La grande scienza; 1987.

[3]   Babbage C. 'On the mathematical powers of the calculating engine'. *The Origins of Digital Computers*. Springer; 1982. pp. 19–54.

[4]   Turing A.M. 'On computable numbers, with an application to the Entscheidungs problem'. *Journal of Mathematics*. 1936;**58**(345–363):5.

[5]   McCulloch W.S., Pitts W. 'A logical calculus of the ideas immanent in nervous activity. 1943'. *The Bulletin of Mathematical Biophysics*. 1990;**52**(4):115–33.

[6]   McCarthy J. *Programs with common sense*. RLE and MIT computation center; 1960.

[7]   Géron A. *Hands-on Machine Learning with Scikit-Learn, Keras, and TensorFlow: concepts, Tools, and Techniques to Build Intelligent Systems*. O'Reilly Media; 2019.

[8]   Müller A.C., Guido S. *Introduction to Machine Learning with Python: A Guide for Data Scientists*. O'Reilly Media Inc; 2016.

[9]   Burkov A. *The Hundred-page Machine Learning Book*. 1. Canada: Andriy Burkov Quebec City; 2019.

[10] Min S., Lee B., Yoon S. 'Deep learning in Bioinformatics'. *Briefings in Bioinformatics*. 2017;**18**(5):851–69.

[11] Baxevanis A.D., Bader G.D., Wishart D.S. *Bioinformatics*. John Wiley and Sons; 2020.

[12] LeCun Y., Bengio Y., Hinton G. 'Deep learning'. *Nature*. 2015;**521**(7553):436–44.

[13] Goodfellow I., Bengio Y., Courville A. *Deep Learning*. MIT Press; 2016.

[14] Ullo S.L., Khare S.K., Bajaj V., Sinha G.R. 'Hybrid computerized method for environmental sound classification'. *IEEE Access*. 2020;**8**:124055–65.

[15] Mitchell T. *Machine Learning*. Inc MHC; 1997.

[16] Hinton G.E., Sejnowski T.J., Poggio T.A., *et al. Unsupervised Learning: foundations of neural computation*. MIT Press; 1999.

[17] Medium. Principal Component Analysis. Available from https://medium.com/apprentice-journal/pca-application-in-machine-learning-4827c07a61db [Accessed August 2020].

[18] Medium. K-means Clustering. Available from https://medium.com/capital-one-tech/k-means-clustering-algorithm-for-machine-learning-d1d7dc5de882 [Accessed August 2020].

[19] Raschka S., Mirjalili V. *Python Machine Learning: Machine Learning and Deep Learning with Python, scikit-learn, and TensorFlow 2*. 3rd Edn. Packt Publishing; 2019. Available from https://books.google.nl/books?id=sKXIDwAAQBAJ.

[20] Krawczyk B. 'Learning from imbalanced data: Open challenges and future directions'. *Progress in Artificial Intelligence*. 2016;**5**(4):221–32.

[21] Phoungphol P., Zhang Y., Zhao Y. 'Robust multiclass classification for learning from imbalanced biomedical data'. *Tsinghua Science and Technology*. 2012;**17**(6):619–28.

[22] Johnson J.M., Khoshgoftaar T.M. 'Survey on deep learning with class imbalance'. *Journal of Big Data*. 2019;**6**(1):27.

[23] Khoshgoftaar T.M., Golawala M., Van Hulse J. 'An empirical study of learning from imbalanced data using random forest'. *19th IEEE International Conference on Tools with Artificial Intelligence (ICTAI 2007)*. **vol. 2**. IEEE; 2007. pp. 310–7.

[24] Tanha J., Abdi Y., Samadi N., Razzaghi N., Asadpour M. 'Boosting methods for multi-class imbalanced data classification: an experimental review'. *Journal of Big Data*. 2020;**7**(1):1–47.

[25] Chawla N.V., Bowyer K.W., Hall L.O., Kegelmeyer W.P. 'SMOTE: synthetic minority over-sampling technique'. *Journal of Artificial Intelligence Research*. 2002;**16**:321–57.

[26] He H., Bai Y., Garcia E.A., *et al.* 'ADASYN: Adaptive synthetic sampling approach for imbalanced learning'. *2008 IEEE International Joint Conference on Neural Networks (IEEE World Congress on Computational Intelligence)*; 2008. pp. 1322–8.

[27] Reichstein M., Camps-Valls G., Stevens B., *et al.* Deep learning and process understanding for data-driven Earth system science. Nature. 2019. 28

Natural Disasters. Available from https://ourworldindata.org/natural-disasters [Accessed September 2020].

[28]   Arinta R.R., EA W.R. 'Natural disaster application on big data and machine learning: a review'. *2019 4th International Conference on Information Technology, Information Systems and Electrical Engineering (ICITISEE)*. IEEE; 2019. pp. 249–54.

[29]   Goswami S., Chakraborty S., Ghosh S., Chakrabarti A., Chakraborty B. 'A review on application of data mining techniques to combat natural disasters'. *Ain Shams Engineering Journal*. 2018;**9**(3):365–78.

[30]   Ullo S.L., Langenkamp M.S., Oikarinen T.P., *et al*. 'Landslide geohazard assessment with convolutional neural networks using sentinel-2 imagery data'. *IGARSS 2019 - 2019 IEEE International Geoscience and Remote Sensing Symposium'*. 2019:9646–9.

[31]   Novellino A., Cesarano M., Cappelletti P., *et al*. 'Slow-motion landslide susceptibility MAP using machine learning techniques and InSAR combining approach: the case study of Termini-Nerano (Italy)'. 2020.

[32]   Infante D., Di Martire D., Di Santolo A.S., *et al*. 'The contribution of spaceborne SAR data in linear infrastructures monitoring activities'. *National Conference of the Researchers of Geotechnical Engineering*. Springer; 2019. pp. 318–27.

[33]   Di Napoli M., Marsiglia P., Di Martire D., Ramondini M., Ullo S.L., Calcaterra D. 'Landslide susceptibility assessment of Wildfire burnt areas through Earth-Observation techniques and a machine Learning-Based approach'. *Remote Sensing*. 2020;**12**(15):2505.

[34]   Carotenuto F., Angrisani A.C., Bakthiari A. 'A new statistical approach for landslide susceptibility assessment in the urban area of Napoli (Italy)'. *Workshop on World Landslide Forum*. Springer; 2017. pp. 881–9.

[35]   Sepe C., Confuorto P., Angrisani A.C. 'Application of a statistical approach to landslide susceptibility map generation in urban settings'. *IAEG/AEG Annual Meeting Proceedings, San Francisco, California, 2018-Volume 1*. Springer; 2019. pp. 155–62.

[36]   Di Napoli M., Carotenuto F., Cevasco A., *et al*. 'Machine learning ensemble modelling as a tool to improve landslide susceptibility mapping reliability'. *Landslides*. 2020;**17**(8):1897–914.

[37]   Allocca V., Coda S., Di Napoli M., *et al*. 'A novel method for groundwater flooding susceptibility mapping induced by groundwater rebound'. *Stoten*. 2020.

[38]   Infante D., Di Martire D., Calcaterra D., Miele P., Scotto di Santolo A., Ramondini M. 'Integrated procedure for monitoring and assessment of linear infrastructures safety (I-Pro MONALISA) affected by slope instability'. *Applied Sciences*. 2019;**9**(24):5535.

[39]   Di Martire D., Paci M., Confuorto P., *et al*. 'A nation-wide system for landslide mapping and risk management in Italy: the second not-ordinary plan of environmental remote sensing'. *International Journal of Applied Earth Observation and Geoinformation*. 2017;**63**(4):143–57.

[40]   ESRI. *Where deep learning meets GIS*. Available from www.esri.com/about/
       newsroom/arcwatch/where-deep-learning-meets-gis/.
[41]   Cohen J.P., HZ L., Lu T. 'Crater detection via convolutional neural net-
       works'. arXiv preprint arXiv:160100978. 2016; Submitted to 47th Lunar and
       Planetary Science Conference (LPSC); 2016.
[42]   Silburt A., Ali-Dib M., Zhu C., *et al.* 'Lunar crater identification via deep
       learning'. *Icarus*. 2019;**317**(6):27–38.
[43]   DeLatte D.M., Crites S.T., Guttenberg N., Tasker E.J., Yairi T. 'Segmentation
       convolutional neural networks for automatic crater detection on Mars'.
       *IEEE Journal of Selected Topics in Applied Earth Observations and Remote
       Sensing*. 2019;**12**(8):2944–57.
[44]   Heras J.M., Yeung K.L., Donati A. *DrMUST: Automating the Anomaly
       Investigation First-Cut*. ESA Special Publication; 2009.
[45]   Martínez-Heras J.A., Donati A., Sousa B. 'DrMUST - a data mining approach
       for anomaly investigation'. *SpaceOps 2012 Conference*; 2012.
[46]   Heras J.M., Donati A. 'Enhanced telemetry monitoring with novelty detec-
       tion'. *AI magazine*. 2014.
[47]   Shi X., Pang J., Liu D. 'Satellite telemetry time series clustering with im-
       proved key points series segmentation'. *2017 Prognostics and System Health
       Management Conference, PHM-Harbin 2017 – Proceedings*; 2017.
[48]   Pang J., Liu D., Peng Y., Peng X. 'Intelligent pattern analysis and anomaly
       detection of satellite telemetry series with improved time series representa-
       tion'. *Journal of Intelligent & Fuzzy Systems*. 2018;**34**(6):3785–98.
[49]   Gong P., Zhang C., Chen M. 'Editorial: deep learning for toxicity and disease
       prediction'. *Frontiers in Genetics*. 2020;**11**:175.
[50]   Jo T., Nho K., Saykin A.J. 'Deep learning in alzheimer's disease: diagnostic
       classification and prognostic prediction using neuroimaging data'. *Frontiers
       in Aging Neuroscience*. 2019;**11**:220.
[51]   Livne M., Rieger J., Aydin O.U., *et al.* 'A U-Net deep learning framework
       for high performance vessel segmentation in patients with cerebrovascular
       disease'. *Frontiers in Neuroscience*. 2019;**13**:97.
[52]   Lin F., Wu Q., Liu J., Wang D., Kong X. 'Path aggregation U-Net model for
       brain tumor segmentation'. *Multimedia Tools and Applications*. 2020;**29**(6).
[53]   Toğaçar M., Ergen B., Cömert Z. 'COVID-19 detection using deep learning
       models to exploit social mimic optimization and structured chest X-ray im-
       ages using fuzzy color and stacking approaches'. *Computers in Biology and
       Medicine*. 2020;**121**:103805.
[54]   Vaishya R., Javaid M., Khan I.H., Haleem A. 'Artificial intelligence (AI)
       applications for COVID-19 pandemic'. *Diabetes & Metabolic Syndrome:
       Clinical Research & Reviews*. 2020;**14**(4):337–9.
[55]   Sebastianelli A., Mauro F., Di Cosmo G., Passarini F., Carminati M., Ullo
       S.L. 'AIRSENSE-TO-ACT: a concept paper for COVID-19 countermeasures
       based on artificial intelligence algorithms and Multi-Source data processing'.
       ISPRS International Journal of Geo-Information. 2020;**10**(1):34.

Chapter 4

Artificial neural network

Maria Pia Del Rosso[1], Silvia Liberata Ullo[1], Alessandro Sebastianelli[1], Dario Spiller[2], Erika Puglisi[3], Filippo Biondi[4], and Danilo Orlando[4]

In this chapter the authors present the theory and mathematics behind artificial neural networks (ANNs) as the starting point for understanding more complex networks that will be presented in the next chapters. This chapter also includes related concepts such as activation functions, gradient descent, optimisers, initialisers, normalisation and dropout, just to list a few.

4.1 Introduction

The original goal of the artificial neural network (ANN) approach was to solve problems in the same way that a human brain does. However, over time, the purpose of ANNs moved to performing specific tasks, including computer vision, speech recognition, machine translation, social network filtering, playing video games and medical diagnosis.

For the mathematical notation, the reader can refer to Appendix A.

4.2 Artificial neural networks

ANNs are computing systems inspired by biological neural networks that constitute the brain of living beings. Such systems 'learn' to perform tasks by considering examples, generally without being programmed with any task-specific rules. For example in image recognition, they might learn to identify images that contain water basins by analysing example images that have been manually labelled as 'water

[1]Engineering Department, University of Sannio, Benevento, Italy
[2]Joint ASI - Φ-Lab, Italian Space Agency (ASI) and European Space Agency (ESA), Rome, Italy
[3]Φ-Lab, European Space Agency (ESA), Frascati, Italy
[4]University Niccolò Cusano, Rome, Italy

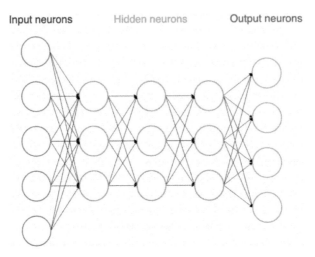

Figure 4.1   *An artificial neural is an interconnected group of nodes, inspired by the simplification of neurons in the brain. Each circle represents an artificial neuron and an arrow represents a connection (i.e. an edge) from the output of one artificial neuron to the input of another*

basin' or 'no water basin' and using the results to identify water basins in new data. This process is called *generalisation*.

An example of an ANN is shown in Figure 4.1: its structure is based on a set of nodes, called *artificial neurons*, connected by *edges* that, like the synapses in a brain, can transmit a signal from one neuron to another. The composition of the input and the output of an artificial neuron will be explained in section 4.2.1. Both the artificial neurons and the edges typically have a weight that changes during the learning/training process and increases or decreases the strength of the signal at a connection. The artificial neurons are also aggregated into *layers*, for example in the figure the input neurons (in blue) constitute the input layer, while the output neurons (in green) constitute the output layer. Between the input and output layers, a user-defined number of hidden intermediate layers can be found. Different layers may perform different transformations on signals which travel through the network from the input layer to the output layer.

The organisation of artificial neurons and edges determines the typology of the network, for example: a neural network with more than one hidden layer is called *deep neural network*; a neural network in which the edges form a directed acyclic graph is called a *feedforward neural network* and a network that allows connections between neurons in the same or previous layers is known as a *recurrent neural network*. Similarly for the layers: a neural network has a *fully connected layer* if every neuron in it is connected to every neuron in the following layer, or has a *pooling layer* if a group of neurons in it connects to a single neuron in the following layer (it is useful to reduce the number of neurons in a layer; for more details in the next chapter we will talk about pooling layer) [1–5].

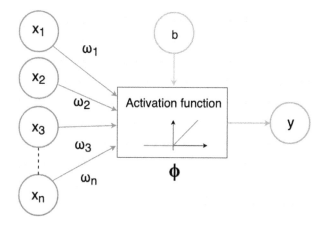

Figure 4.2   Example of an artificial neuron

4.2.1   *Artificial neuron*

Artificial neurons are elementary units in ANNs. The classic model of an artificial neuron is represented in Figure 4.2: it receives one or more inputs x_i, separately weighted with weights ω_i, in order to perform a weighted summation whose result is passed through a non-linear *activation function* to produce the output y. The y value can either pass through other neurons or represent the result of the model, depending on the architecture of the network.

The mathematical function that describes the behaviour of the artificial neuron is called *propagation function* and is given by

$$z = x_1 \cdot \omega_1 + x_2 \cdot \omega_2 + x_3 \cdot \omega_3 + \cdots + x_n \omega_n + b = x^T \omega + b \qquad (4.1)$$
$$y = \Phi(z) \qquad (4.2)$$

where x_i is the i-th component of the input vector x, ω_i is the i-th weight in weights vector ω, b is the bias neuron and Φ is the activation function discussed in section 4.3. The bias neuron is a special neuron, added to each layer in the neural network, which simply stores a value and acts like a constant which helps the model to fit the given data [1–5]. A practical example of its utility is given in section 4.3.4.

4.3   Activation functions

In the ANN, activation functions are mathematical equations that determine whether a node should be activated ('fired') or not (based on its relevance in the model's prediction) and the output of that node given a set of inputs. Moreover, they help to normalise the output of each neuron to a specific range that depends on the typology of the applied activation function (the most common typologies will be explored in the next subsections). There exist three principal types of activation functions:

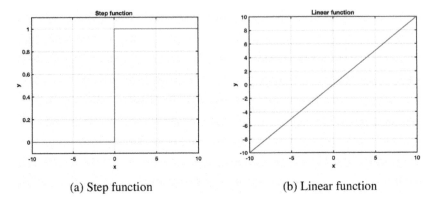

(a) Step function   (b) Linear function

Figure 4.3   Step and linear activation functions

- *Step function*: is a threshold-based activation function (see Figure 4.3). If the input value is above a certain threshold, the neuron is activated and sends a1 to the next layer, otherwise, the output is 0. This kind of function could be used to implement a binary classifier that only needs to answer 'yes' or 'no' (e.g. 'activate' or 'not activate') to specific requests.
- *Linear function*: is a first degree polynomial in which the output signal is proportional to the input (see Figure 4.3). Even if it is better than a step function because the output can take more values in $\mathbb{R}$, it is still limited in its complexity and it is not effective to learn complex functional mappings from data such as images, videos, audio, speech, etc. Practically, no matter how many layers are implemented in the neural network, the output signal will be a simple linear function of the first layer (because a linear combination of linear functions is still a linear function), so it is like there is no activation function applied to the node. This is the working method of linear regression models. Moreover, back propagation (analysed in section 4.4) cannot be applied because the derivative of the function is a constant and has no explicit relation to the input. Hence, it is not possible to understand which weights in the input neurons can provide better prediction.
- *Non-linear function*: modern neural networks use this kind of functions to learn complex data and provide accurate predictions. Using a non-linear activation function, non-linear mappings from inputs to outputs are feasible. Another important quality of these activation functions is that they are differentiable, thus the back propagation strategy can be applied to reduce errors.

The most common non-linear activation functions are presented in the following subsections.

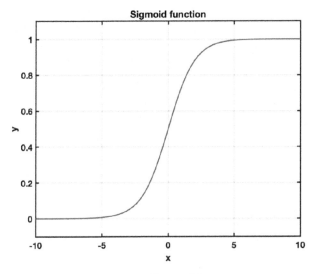

Figure 4.4   Sigmoid function

4.3.1   Sigmoid function

The sigmoid function is given as

$$\Phi(z) = \frac{1}{1 + e^{-z}}. \tag{4.3}$$

It is a function with an S-shaped graph (see Figure 4.4) ranging between 0 and 1 in the y coordinate and having a smooth gradient. It is easy to understand and apply, but its popularity has fallen due to three major reasons:

- Vanishing gradient problem; this problem will be presented in subsection 4.7.
- Its output is not zero-centred; this makes the optimisation harder.
- It is computationally expensive.

4.3.2   Hyperbolic tangent function (tanh)

The hyperbolic tangent function activation is computed as:

$$\Phi(z) = \tanh(z) = \frac{1 - e^{-2z}}{1 + e^{-2z}}. \tag{4.4}$$

The output is centred at zero; this is because the range sampling is done with $-1 < 1$ (see Figure 4.5). When the optimisation strategy is determined using a rectified linear unit function (ReLU), the algorithm is generally simpler and therefore preferable to a sigmoid function. However, it still suffers from the problem quantified as disappearance gradient.

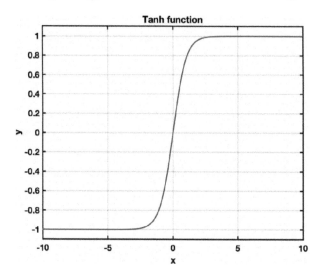

Figure 4.5   Tanh function

4.3.3   *Rectified linear unit function*

'ReLU' has become very popular in recent years because it is very efficient from the point of view of computational complexity. Indeed, due to its local linearity characteristic, it converges very quickly and is a good approximation of more complex functions (in fact, any function can be approximated as combinations of several ReLu). Its mathematical formula is

$$\Phi(z) = \max(0, z). \tag{4.5}$$

This function avoids and corrects the 'vanishing gradient' problem, so it is less computationally expensive compared with the tanh and sigmoid functions that involve the calculation of mathematical operations, are of a non-linear nature and are certainly more complex.

Also, ReLU has a limitation: since it can suffer from the dying ReLU problem, it should only be used within the hidden layers of an ANN. In fact, because of the horizontal line in ReLU (as can be seen in Figure 4.6), the gradient of the function becomes zero when the inputs approach zero, or are negative, and the network cannot perform back propagation and cannot learn from the data. Having 'dead neurons' in the output layer means the output is equal to zero. To fix the problem of dying neurons, a small slope in the negative area was introduced to enable back propagation even for negative values of the input. The function containing this variation is called the leaky ReLU function (see Figure 4.7) and the mathematical formula is:

$$\Phi(z) = \max(0.01z, z) \tag{4.6}$$

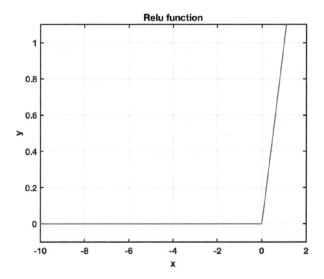

Figure 4.6   The rectified linear unit function (ReLU) function

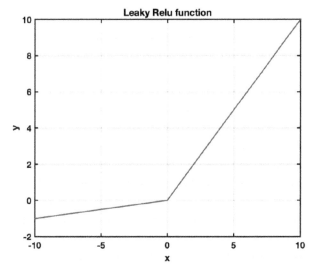

Figure 4.7   The leaky rectified linear unit function (ReLU) function

4.3.4   The bias

The bias is an additional parameter in neural networks that allows shifting of the activation function to either the right or the left on the graph. The best way to understand its utility is to report an example of a one-input one-output neural network (Figure 4.8).

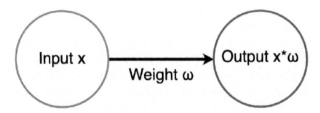

Figure 4.8   A one-input one-output neural network

The output of the network is computed by multiplying the input x by the weight ω and passing the result through an activation function. If a sigmoid function is chosen, then

$$y = \frac{1}{1+e^{-x\cdot\omega}}. \tag{4.7}$$

In Figure 4.9, the output of the network is computed using different values of ω (e.g. $\omega \in \{0.5, 1.0, 2.0\}$). Essentially, to change the weight ω means to change the steepness of the sigmoid function. However, without using the bias, the three curves in Figure 4.9 are naturally constrained to pass through the point $[0, 0.5]$. Consequently, the network would never reach, for instance, an output with a value of 0 taking 2 as the input value. This result can be reached introducing a bias to move to the right the curve in the graph. Equation (4.7) must be changed as

$$y = \frac{1}{1+e^{-x\cdot\omega+b}}. \tag{4.8}$$

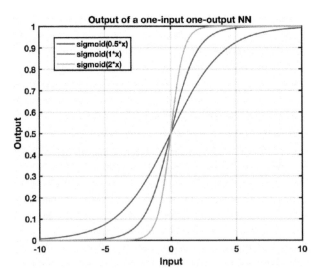

Figure 4.9   Graphic representation of results of a one-input one-output neural network without bias with different values of the weight w, using the sigmoid as the activation function

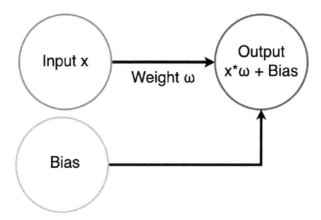

Figure 4.10   *A one-input one-output neural network with a bias neuron*

The structure of the neural network necessary to solve a general problem is repre-sented in Figure 4.10, while Figure 4.11 shows the results obtained using different values of bias and a fixed value of weight w. The red curve has no bias, the blue curve has a bias value of -5 (i.e. a negative bias shifts the curve to the left) and the yellow curve has a bias value of $+5$ (i.e. a positive bias shifts the curve to the right). Figure 4.11 shows that with a bias value of $+5$ an output value of 0 can be achieved with 2 as the input value. In conclusion, a biased neuron can learn such input vectors

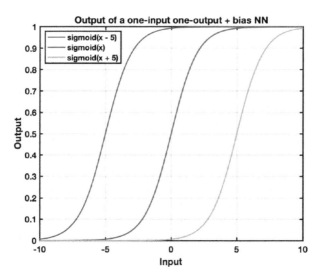

Figure 4.11   *Graphic representation of results of a one-input one-output neural network with different values of bias and fixed weight w, using the sigmoid as the activation function*

that a non-biased neuron is not able to learn. The additional weight costs more calculations, but is also necessary to improve neuron properties.

4.4   Forward and backward propagation

During the training phase of an ANN, the signal is forward propagated to get the output $\mathbf{y}(\boldsymbol{\omega})$ that is compared with the real value y_{true} to get the error evaluated by means of an objective function $J(\mathbf{x}; y_{true})$. As an example, supposing an output y of size N, the error can be associated to a mean square error function,

$$J(\boldsymbol{\omega}; \mathbf{y}_{true}) = \frac{\sum_{i=1}^{N}(y_i(\boldsymbol{\omega})-y_{i,true})^2}{N} \qquad (4.9)$$

or to a mean absolute error

$$J(\boldsymbol{\omega}; \mathbf{y}_{true}) = \frac{\sum_{i=1}^{N}|y_i(\boldsymbol{\omega})-y_{i,true}|}{N} \qquad (4.10)$$

To minimise the error, the signal is then propagated backwards to find the derivative of the error (with respect to each weight) that needs to be subtracted from the weight value. This update is done using a form of gradient descent. Then, a new forward propagation is computed to see how well the new weights are performing and, again, follows a backward propagation to update the weights. This will go on until a minimum value for the error is reached. The learning process of ANNs deals with understanding the number of neurons to activate and the parameters of the activation, depending on the type of inputs received.

4.4.1   The gradient descent

During the training phase, a machine learning model must complete an optimisation process to - 'learn' information from the data, typically driven by the gradient descent technique.

Gradient descent is a first-order iterative optimisation algorithm for finding the minimum of a function. To find a local minimum of a function using the gradient descent, we proceed proportionally to the negative of the gradient of the function at the current point.

Gradient descent is an optimisation algorithm often used for finding the weights or coefficients of machine learning algorithms, such as ANNs. Its operation requires that the model to be trained makes previsions on the training data and that the error made by the model on the previsions is used to update the weights of the network, reducing the error. The goal is to find the parameters of the model (e.g. weights) that minimise the prevision error on the training dataset. This result is achieved by changing the model parameters so that the output moves along the gradient or slope of the error to reach a minimum error value. This gives the algorithm its name of 'gradient descent' [6–9].

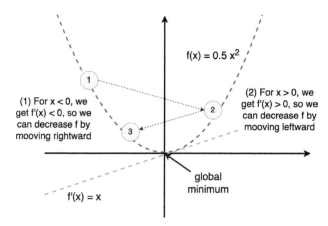

$f(x) = 0.5 \; x^2$

(1) For x < 0, we get f'(x) < 0, so we can decrease f by mooving rightward

(2) For x > 0, we get f'(x) > 0, so we can decrease f by mooving leftward

global minimum

f'(x) = x

Figure 4.12   An illustration of how the gradient descent algorithm uses the derivatives of a function to reach a minimum

4.4.2   Description

Most deep learning algorithms involve optimisation of some sort. Without loss of generality, let $f(x) : \mathbb{R} \to \mathbb{R}$ represents the aforementioned objective function J. Optimisation refers to the task of either minimising or maximising some function $f(x)$ by modifying x, even though most optimisation problems are cast in terms of minimising $f(x)$. Maximisation may be accomplished via a minimisation algorithm by considering $-f(x)$.

The function we want to minimise or maximise is called the objective function but, in case of minimisation, it can also be referred to as cost function, loss function or error function. Considering a scalar input $x \in \mathbb{R}$, the derivative $f'(x)$ gives the slope of $f(x)$ at the point x. In other words, it specifies the first order approximation to map a small change in the input to the corresponding change in the output: $f(x + \varepsilon) \simeq f(x) + \varepsilon f'(x)$. The derivative is therefore useful for minimising a function because it provides information on how to change x in order to make a small improvement in y. From basic analysis, it is known that $f(x - \varepsilon sign(f'(x)))$ is less than $f(x)$ for small enough ε. Hence, $f(x)$ can be reduced by moving x in small steps in the opposite direction of the derivative. This technique is called gradient descent. See Figure 4.12 for an example of this technique.

When $f'(x) = 0$, the derivative provides no information about which direction to move in. Points where $f'(x) = 0$ are known as critical or stationary points. A point that obtains the absolute lowest value of $f(x)$ is a global minimum. It is possible that the function has only one global minimum or multiple global minimums. It is also possible that the function has several local minima, which are not globally optimal. In the context of deep learning, it is often necessary to optimise functions that have many suboptimal global minimums, many saddle points surrounded by flat regions. This makes the optimisation process very difficult, especially when the input of the function is very complex and multidimensional. Usually, finding a very low value of

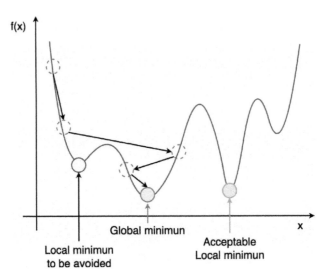

Figure 4.13   *Optimisation algorithms may not find a global minimum when the function returns multiple local minimum points. In the context of deep learning, we generally accept such solutions even if they do not belong to the absolute minima. This approximation is valid as long as solutions correspond to significantly low values of the cost function (these values should be comparable with the absolute minima)*

$f(x)$ is a feasible and commonly accepted solution, even though it is not a minimal point in any formal sense. See Figure 4.13 for an example.

For many real-world problems, the minimisation is performed over functions that have multiple inputs and one scalar output, i.e. $f(x) : \mathbb{R}^n \Rightarrow \mathbb{R}$. It is noteworthy that, for the concept of 'minimisation' to make sense, there must still be only one scalar output. For functions with multiple inputs, the aforementioned derivative is substituted by the gradient, a vector collecting all the partial derivatives of $f(x)$ and denoted as $\nabla_x f(x)$. The i-th element of the gradient is the partial derivative of f with respect to x_i. The partial derivative $\frac{\partial f(x)}{\partial x_i}$ measures how f changes as only the variable x_i increases at point x. The gradient generalises the notion of derivative to the case where the derivative is with respect to a vector. In multiple dimensions, critical points are points where every element of the gradient is equal to zero.

The directional derivative in direction $\hat{u}$ (a unit vector) is the slope of the function f in direction $\hat{u}$. In other words, the directional derivative is the derivative of the function $f(x + \alpha \hat{u})$ with respect to $\alpha \in \mathbb{R}$, evaluated with $\alpha \to 0$. Using the chain rule, it can be shown that $\frac{\partial f(x + \alpha \hat{u})}{\partial \alpha}$ is equal to $\hat{u}^T \nabla_x f(x)$ when $\alpha \to 0$.

To minimise f, the direction in which f decreases the fastest must be searched for. With this aim, the directional derivative is used as

$$\min_{\hat{u}} \hat{u}^{\mathrm{T}} \nabla_x f(x) = \min_{\hat{u}} \|\hat{u}\|_2 \|\nabla_x f(x)\|_2 \cos\theta, \tag{4.11}$$

where θ is the angle between u and the gradient. By definition $\|\hat{u}\|_2 = 1$, and ignoring factors that do not depend on $\hat{u}$, (4.11) simplifies to $\min_u \cos\theta$. This is minimised when $\hat{u}$ points in the opposite direction with respect to the gradient. In other words, the gradient points directly uphill, and the negative gradient points directly downhill. Hence, f is decreased by moving in the direction of the negative gradient. This is known as the method of gradient descent.

Gradient descent is based on the evaluation of new points by using the criterion

$$x' = x - \varepsilon \nabla_x f(x) \tag{4.12}$$

where $\varepsilon \in \mathbb{R}$ is the learning rate, a positive scalar determining the size of the step. The parameter ε can be chosen in several different ways. A common approach is to set it to a small constant. Sometimes, the problem is solved using the step size that makes the directional derivative vanish. Another approach is to evaluate $f(x-\varepsilon \nabla_x f(x))$ for several values of ε and choose the one that results in the smallest objective function value. This last strategy is called a *line search*.

Gradient descent converges when every element of the gradient is zero, in practice, when every element of the gradient is very close to zero. In some cases, analytical solutions can be found, thus avoiding running the iterative algorithm and directly jumping to the critical point by solving the equation $\nabla_x f(x) = 0$ for x.

4.4.3   *Stochastic gradient descent*

Stochastic gradient descent (SGD) is a variation of the gradient descent algorithm that calculates the error and updates the model for each example in the training dataset [10–13].

Pros and cons of SGD
Pros

- The frequent updates immediately give an insight into the performance of the model and the rate of improvement.
- This variant of gradient descent may be the simplest to understand and implement, especially for beginners.
- The increased model update frequency can result in faster learning for some problems.
- The noisy update process can allow the model to avoid local minima (e.g. premature convergence).

Cons

- Updating the model so frequently is more computationally expensive than other configurations of gradient descent, taking significantly longer to train models on large datasets.

- The frequent updates can result in a noisy gradient signal, which may cause the model parameters and in turn the model error to jump around (have a higher variance over training epochs).
- The noisy learning process down the error gradient can also make it hard for the algorithm to settle on an error minimum for the model.

4.4.4   Full batch gradient descent

Batch gradient descent is a variation of the gradient descent algorithm that calculates the error for each example in the training dataset, but only updates the model after all training examples have been evaluated. The average of the gradients of all the training examples is considered and then that mean gradient is used to update the model parameters.

One cycle through the entire training dataset is called a training epoch. Therefore, it is often said that batch gradient descent performs model updates at the end of each training epoch.

Pros and cons of full batch gradient descent
Pros

- Fewer updates to the model means this variant of gradient descent is more computationally efficient than SGD.
- The decreased update frequency results in a more stable error gradient and may result in a more stable convergence in some problems.
- The graph of cost versus epochs is quite smooth.
- The separation of the calculation of prediction errors and the model update lends the algorithm to parallel processing based implementations.

Cons

- The more stable error gradient may result in premature convergence of the model to a less optimal set of parameters.
- The updates at the end of the training epoch require the additional complexity of accumulating prediction errors across all training examples.
- Commonly, batch gradient descent is implemented in such a way that it requires the entire training dataset loaded in memory and ready to use for the algorithm.
- Model updates, and in turn training speed, may become very slow for large datasets.

4.4.5   Mini-batch gradient descent

Mini-batch gradient descent is a variation of the gradient descent algorithm that splits the training dataset into small batches that are used to calculate model error and update model coefficients. Implementations may choose to sum the gradient over the mini-batch which further reduces the variance of the gradient.

Mini-batch gradient descent seeks to find a balance between the robustness of SGD and the efficiency of batch gradient descent. It is the most common implementation of gradient descent used in the field of deep learning [14].

Pros and cons of mini-batch gradient descent

Pros

- The model update frequency is higher than batch gradient descent which allows for a more robust convergence, avoiding local minima.
- The batched updates provide a computationally more efficient process than SGD.
- The batching allows both the efficiency of not having all training data in memory and algorithm implementations.

Cons

- Mini-batch requires the configuration of an additional 'mini-batch size' hyperparameter for the learning algorithm.
- Error information must be accumulated across mini-batches of training examples like batch gradient descent.

Mini-batch gradient descent is the recommended variant of gradient descent for most applications, especially in deep learning. Mini-batch sizes, commonly called 'batch sizes' for brevity, are often tuned to an aspect of the computational architecture on which the implementation is being executed. Such as a power of 2 that fits the memory requirements of the GPU (Graphics Processing Unit) or CPU (Central Processing Unit) hardware like 32, 64, 128, 256 and so on. Batch size is a slider on the learning process:

- Small values give a learning process that converges quickly at the cost of noise in the training process.
- Large values give a learning process that converges slowly with accurate estimates of the error gradient.

4.4.5.1 Tips
- A good default for batch size might be 32 [15].
- It is a good idea to review learning curves of model validation error against training time with different batch sizes when tuning the batch size.
- Tune batch size and learning rate after tuning all other hyperparameters.

4.4.6 *Batch normalisation*

To increase the stability of a neural network, batch normalisation normalises the output of a previous activation layer by subtracting the batch mean and dividing by the batch standard deviation. However, after this shift/scale of activation outputs by some randomly initialised parameters, the weights in the next layer are no longer optimal. SGD undoes this normalisation if it's a way for it to minimise the loss function. Consequently, batch normalisation adds two trainable parameters to each layer, so the normalised output is multiplied by a 'standard deviation' parameter γ and adds a 'mean' parameter γ. In other words, batch normalisation lets SGD do the denormalisation by changing only these two weights for each activation, instead of losing the stability of the network by changing all the weights.

4.4.7   *The vanishing gradient*

The vanishing gradient is a problem that leads to a decrease in the gradient so as to block learning. As more layers are added to neural networks, the gradients of the loss function approach 0, making the network hard to train. Certain activation functions, like the sigmoid function, squish a large input space into a small output space between 0 and 1. Therefore, a large change in the input of the sigmoid function will cause a small change in the output. Hence, the derivative becomes small.

For a shallow network with only a few layers that use these activations, this is not a big problem. However, when more layers are used, it can cause the gradient to be too small for training to work effectively. Gradients of neural networks are found using back propagation. Simply put, back propagation finds the derivatives of the network by moving layer by layer from the final layer to the initial one. By the chain rule, the derivatives of each layer are multiplied down the network (from the final layer to the initial) to compute the derivatives of the initial layers. However, when hidden layers use an activation like the sigmoid function, n small derivatives are multiplied together. Thus, the gradient decreases exponentially as we propagate down to the initial layers. A small gradient means that the weights and biases of the initial layers will not be updated effectively with each training session. Since these initial layers are often crucial for recognising the core elements of the input data, it can lead to overall inaccuracy of the whole network [16–18].

4.4.7.1   Solutions

The simplest solution is to carefully choose the activation function. For instance, the ReLU does not usually lead to the presence of small derivatives.

Residual networks [19] are another solution, as they provide residual connections straight to earlier layers. This residual connection does not go through activation functions that 'squashes' the derivatives, resulting in a higher overall derivative of the block.

Finally, batch normalisation layers can also resolve the issue. As stated before, the problem arises when a large input space is mapped to a small one, causing the derivatives to disappear. Batch normalisation reduces this problem by simply normalising the input so, for example $|x|$ does not reach the outer edges of the sigmoid function.

4.4.8   The gradient explosion

Exploding gradients are a problem where large error gradients accumulate and result in very large updates to neural network model weights during training. This has the effect of your model being unstable and unable to learn from the training data.

In deep networks or recurrent neural networks, error gradients can accumulate during an update and result in very large gradients. As a consequence, the training leads to unstable networks. Eventually, the values of weights can become so large as to overflow and result in NaN (Not a Number) values. The explosion occurs through exponential growth by repeatedly multiplying gradients through the network layers that have values larger than 1.0. There are some subtle signs that you may be suffering from exploding gradients during the training of your network, such as:

- The model is unable to get traction on your training data (e.g. poor loss).
- The model is unstable, resulting in large changes in loss from update to update.
- The model loss goes to NaN during training.

4.4.8.1   Solution

There are many approaches to address exploding gradients; this section lists some best practice approaches that you can use [20–22].

- Re-design the network model (e.g. modify the number of layers, modify the types of layers in the network, etc.).
- Use gradient clipping to place a predefined threshold on the gradients to prevent their norm from getting too large. Specifically, the values of the error gradient are checked against a threshold value and clipped or set to that threshold if they exceed.
- Use weight regularisation to check how big the network weights are and add a penalty to the network loss function to minimise the risk of having large weight values.

4.4.9   The back propagation

A supervised neural network, at the simplest abstract representation, can be presented as a box with two functioning modalities *learning* and *prediction* as shown in Figure 4.14.

The learning process uses inputs and desired outputs, both known at this stage, and updates its internal states using them. The goal is to have the computed output as close as possible to the desired output. The prediction process uses new inputs to generate an output, using the internal state trained before.

Back propagation algorithms are a family of methods used to efficiently train ANNs following a gradient-based optimisation algorithm. The main feature of back propagation is its iterative, recursive and efficient method for calculating the weights

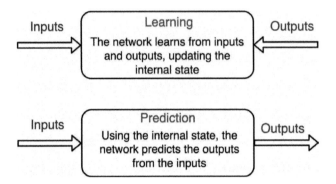

Figure 4.14   Neural network as box

updates to improve the network until it is able to perform the task for which it is being trained.

Back propagation requires the derivatives of activation functions to be known at network design time. Automatic differentiation is a technique that can automatically and analytically provide the derivatives to the training algorithm. In the context of learning, back propagation is commonly used by the gradient descent optimisation algorithm to adjust the weight of neurons by calculating the gradient of the loss function; back propagation computes the gradient(s), whereas (stochastic) gradient descent uses the gradients for training the model (via optimisation).

4.4.10   A numerical example

In this section, the authors present a numerical example of how back propagation works. The steps are summarised in Figure 4.15 and the reader can find a scheme describing the learning process.

An easy way for understanding neural networks and supervised learning is to start with a numerical example and a one-input and one-output model, with a linear

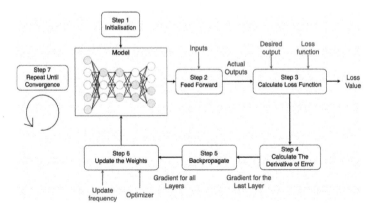

Figure 4.15   Neural networks step by step

Table 4.1   Back propagation: forward propagate after initialisation

Input	0	1	2	3	4
Desired output	0	2	4	6	8

Table 4.2   Back propagation: forward propagation after initialisation

Input	0	1	2	3	4
Model output with $\omega = 3$	0	3	6	9	12

relation. The goal is to search which function fits better the date among all the possible linear functions. Table 4.1 shows the dataset for the abovementioned example considering that the relationship between the input and the desired output is given as

$$\text{desired output} = 2 \cdot \text{input} \tag{4.13}$$

However, for most of the real cases the relationship between the input and output is far more complex and highly non-linear.

Step 1 – Model initialisation

The first step for the learning process is to start with an initial hypothesis. A random initialisation of the model is a common practice. The reason behind this choice is that whatever the initialisation, through an iterative learning process, the pseudo-ideal model can be obtained. In this example, the goal is to find which model, with the generic form expressed by 4.14, can fit at the best the dataset.

$$y = \omega \cdot x \tag{4.14}$$

In (4.14), ω is the weight of the network and can be initialised randomly. These types of models are simply called feed-forward linear layers.

Step 2 – Forward propagate

After initialising the model, the natural step is to check its performance. Assuming, for example, that the random initialisation leads to $\omega = 3$, the actual output can be calculated (see Table 4.2).

This step is called forward propagation, because the calculation flow is going in the natural forward direction from the input, through the neural network and to the output.

Step 3 – Loss function

After the forward propagation, the model performances must be evaluated. To this aim, the loss function is introduced to measure the difference between the known input label and the predicted label. The most intuitive loss function is given as

$$loss = (\text{desired output} - \text{actual output}) \tag{4.15}$$

Table 4.3   Back propagation: loss functions

Input value	Current value	Desired value	Absolute error	Square error
0	0	0	0	0
1	3	2	1	1
2	6	4	2	4
3	9	6	3	9
4	12	8	4	16
Total	-	-	10	30

However this loss function returns positive values when the network undershoots (prediction < desired output), and negative values when the network overshoots (prediction > desired output). For these reasons a better loss function can be the one expressed by (4.16).

$$loss = |(\text{desired output} - \text{actual output})| \tag{4.16}$$

However, several situations can lead to the same total sum of errors: for instance, a lot of small errors or few big errors can sum up exactly to the same total amount of error. In order to allow the NN (Neural Network) to converge to such a situation, the most famous loss function in NN can be used, expressed by (4.17).

$$loss = \sum |(\text{desired output} - \text{actual output})|^2 \tag{4.17}$$

The calculations of the loss values for the current example are shown in Table 4.3.

Note that if you consider only the first input, it appears that the network is able to correctly predict the output. To solve that problem, the overall error over the whole dataset has to be minimised. Indeed the loss function is an error metric, that gives an indicator on how much precision gets lost, if the real desired output is replaced with the actual output generated by the trained neural network model. The machine learning goal then would be to minimise the loss function so it can be transformed into an optimisation process that aims to minimise the loss function.

Step 4 – Differentiation

Any optimisation technique that modifies the internal weights of neural networks can be used. In this simple numerical example, with only one parameter to optimise (ω), you can search from -100.0 to $+100.0$ with a step of 0.001, in which ω has the smallest sum of squares of errors over the dataset. This can work if the model has very few parameters only. However, in real cases the models have millions of weights to optimise and a brute force approach cannot be imaginable, since it is a pure waste of computational resources.

A typical method used to solve this problem, as discussed in section 4.1, is the gradient descent, that involves the use of derivatives. Basically the idea is to evaluate the model when the internal state is varied by a small value $\delta\omega$. Assuming

Table 4.4   Back propagation: loss functions

Input	Output	$\omega = 3$	rmse (3)	$\omega = 3.0001$	rmse
0	0	0	0	0	0
1	2	3	1	3.0001	1.0002
2	4	6	4	6.0002	4.0008
3	6	9	9	9.0003	9.0018
4	8	12	16	12.0004	16.0032
Total	-	-	30	-	30.006

$\delta\omega = 0.0001$, Table 4.4 shows the recalculated errors (where RMSE stands for Root Mean Square Error).

As you can see from Table 4.4, by increasing ω from 3 to 3.0001, the sum of squares of error will increase from 30 to 30.006. Since the best function that fits this model is $y = 2 \cdot x$, increasing the weights from 3 to 3.0001 should increase the error a little bit more.

Here is what the loss function looks like:

- If $\omega = 2$, the loss is 0, since actual output of the neural network will perfectly fit the training set.
- If $\omega < 2$, the loss function assumes positive values, but the derivative is negative, meaning that an increase in weight will decrease the loss function.
- At $\omega = 2$, the loss is 0 and the derivative is 0, a perfect model is reached, nothing is needed.
- If $\omega > 2$, the loss becomes positive again, but the derivative is positive as well, meaning that any more increase in weight will increase the losses even more.

Step 5 – Back propagation

The derivative is decomposable, thus it can be back propagated. This can be done by de-stacking through the function calls. This technique is called auto-differentiation, and only requires each function to be provided with the implementation of its derivative.

Step 6 – Weight update

As shown previously, the derivative is just the rate at which the error changes relatively to the change in weight. In the current numerical example this rate is 60 ×. This means that 1 unit of change in weight leads to 60 units change in error. Knowing that the error is currently at 30 units, extrapolating the rate in order to reduce the error to 0, it is necessary to reduce the weights by 0.5 units. However, in real problems weights are not updated in such large steps. Since there are many non-linearities, any major changes in weights will lead to chaotic behaviour. We must not forget that the derivative is local only at the point where we are calculating the derivative. As a general rule for weight updates you can use the delta rule, expressed by (4.18):

$$\text{New weight} = \text{old weight} - \text{derivative rate} \cdot \text{learning rate} \tag{4.18}$$

The learning rate, in this case, is a constant, in order to force the weight to get updated very smoothly and slowly (to avoid big steps and chaotic behaviour). Several weight update methods, called optimisers, exist. The delta rule is the most simple and intuitive one, however it has several drawbacks.

Step 7 – Iterate until convergence

Since the weights are updated a small delta step at a time, it will take several iterations in order to learn. In neural networks, after each iteration, the gradient descent force updates the weights. The number of iterations needed to reach the convergence depends on different factors:

- On how strong is the learning rate applied. A high learning rate means faster learning, but with more chances of instability.
- On the meta-parameters of the network (how many layers, how complex the non-linear functions are). The more the number of variables it has the longer it takes to converge, and the higher the precision it can reach.
- It depends on the optimisation method used; some weight update rules are proven to be faster than others.
- On the random initialisation of the network.
- On the quality of the training set. If there is no correlation between the input and the output, the neural network will not be able to learn a random correlation.

4.5   Overfitting and underfitting

Overfitting and underfitting are very important basic concepts when dealing with the training of machine learning models including neural networks. In a nutshell, overfitting refers to the case where the model fits the data in the training dataset quite well but fails to predict the data in the test dataset. On the contrary, underfitting happens when the model does not learn enough information during the training phase and poorly predicts new data.

4.5.1   Bias and variance

Bias and variance are two important statistical concepts for understanding overfitting and underfitting.

- The **bias** is a value indicating how different our predictions are from the correct values. High bias can cause an algorithm to miss the relevant relations between features and target outputs (underfitting)
- The **variance** is a value indicating the difference in the predictions using different parts of the input dataset. High variance can cause an algorithm to learn noise in the training data, rather than the intended outputs (overfitting).

To sum up, a model with low bias and high variance is a model with overfitting, while a model with high bias and low variance is usually a model with underfitting. The algorithm reaches the best results when both bias and variance are low.

Causes of overfitting can be:

- An insufficient number of examples in the training dataset
- A higher number of features than the number of examples
- A highly complex network.

The remedies for overfitting can be:

- To increase the number of examples in the dataset
- To decrease the number of features
- To reduce the network complexity
- To use a regularisation technique (see section 4.6).

The causes of underfitting can be:

- The training process stopped too early
- The model is excessively simple.

In this case, the solution can be reviewing the training phase or adding a little bit of complexity to the model.

4.5.2   *Early stopping of training*

When training a large model on a sufficiently large dataset, if the training is done for a long amount of time rather than increasing the generalisation capability of the model, it increases the overfitting. As in the training process, the training error keeps on reducing but after a certain point, the validation error starts to increase, signifying that our model has started to overfit (see Figure 4.16).

Early stopping can be associeted to a very efficient hyperparameter selection algorithm. The idea of early stopping of training is that as soon as the validation error starts to increase we freeze the parameters and stop the training process. Or we can also store the copy of model parameters every time the error on the validation set improves and return these parameters when the training terminates rather than the latest parameters. The advantage of early stopping over weight decay is that it auto-matically determines the correct amount of regularisation, whereas weight decay requires many training experiments with different values of its hyperparameters.

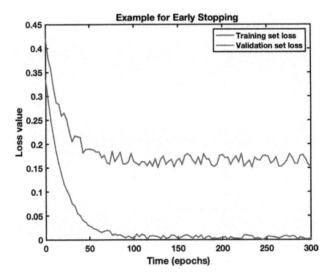

Figure 4.16   Early stopping example

4.5.3   Dropout

Dropout can be compared as a method of making bagging[a] practical for ensembles of very large neural networks. The method of bagging cannot be directly applied to large neural networks as it involves training multiple models, and evaluating multiple models on each test example. Since training and evaluating such networks is costly in terms of runtime and memory, this method is impractical for neural networks. Dropout provides an inexpensive approximation to training and evaluating a bagged ensemble of many neural networks. Dropout trains the ensemble consisting of all subnetworks that can be formed by removing non-output units from an underlying base network. In most modern neural networks, based on a series of affine transformations and non-linearities, we can effectively remove a unit from a network by multiplying its output value by zero. This procedure requires some slight modification for models such as radial basis function networks, which take the difference between the unit's state and some reference value. Here, we present the dropout algorithm in terms of multiplication by zero for simplicity, but it can be trivially modified to work with other operations that remove a unit from the network. Dropout training is not quite the same as bagging training. In the case of bagging, the models are all independent. In the case of dropout, the models share parameters, with each model inheriting a different subset of parameters from the parent neural network. This parameter sharing makes it possible to represent an exponential num-

---

[a]Bagging is a machine learning technique that falls into the category of ensemble learning. In bagging several models of the same type are trained on different datasets, each obtained from the initial dataset through random sampling with replacement (bootstrap). The name bagging derives from the combination of the English words bootstrap (i.e. random sampling with replacement) and aggregation (in reference to the aggregation of multiple models, typical of ensemble learning).

ber of models with a tractable amount of memory. One advantage of dropout is that it is very computationally cheap. Using dropout during training requires only $O(n)$ computation per example per update, to generate n random binary numbers and multiply them by the state. Another significant advantage of dropout is that it does not significantly limit the type of model or training procedure that can be used. It works well with nearly any model that uses a distributed representation and can be trained with SGD.

4.6 Regularisation techniques

Strategies or techniques which are used to reduce the error on the test set at an expense of increased training error are collectively known as regularisation. Regularisation can be defined as any modification we make to a learning algorithm that is intended to reduce its generalisation error but not its training error. This regularisation is often done by putting some extra constraints on a machine learning model, such as adding restrictions on the parameter values or by adding extra terms in the objective function that can be thought of as corresponding to a soft constraint on the parameter values. If chosen correctly these can lead to a reduced testing error. An effective regulariser is said to be the one that makes a profitable trade-off by reducing significantly the variance while not overly increasing the bias.

4.6.1 Parameter norm penalties

This technique is based on limiting the capacity of models, by adding a parameter norm penalty to the objective function J.

$$\widetilde{J}(\omega; y_{true}) = J(\omega; y_{true}) + \alpha \Gamma(\omega), \tag{4.19}$$

where α is a hyperparameter that weighs the relative contribution of the penalty function. Setting α to 0 means no regularisation (so that (4.19) comes back to something like (4.9) or (4.10)) while larger values of α correspond to more regularisation. For neural networks, we choose to use parameter norm penalty that penalises on the weights of the affine transformations and leave biases unregularised. This is because of the fact that biases require lesser data to fit accurately than the weights. Weights are used to denote the relationship between the two variables and require observing both variables in various conditions whereas bias controls only single variables, hence they can be left unregularised.

4.6.1.1 Regularisation L2

This regularisation is popularly known as weight decay. This strategy drives the weights closer to the origin by adding the regularisation term ω which is defined as

$$\Gamma(\omega) = \tfrac{1}{2}\|\omega\|_2^2 = \tfrac{1}{2}\omega^\mathsf{T}\omega \tag{4.20}$$

The objective function after regularisation is denoted by

$$\widetilde{J}(\omega;y_{true}) = \tfrac{\alpha}{2}\omega^{\mathsf{T}}\omega + J(\omega;y_{true}), \tag{4.21}$$

and the corresponding gradient is

$$\nabla_{\omega}\widetilde{J}(\omega;y_{true}) = \alpha\omega + \nabla_{\omega}J(\omega;y_{true}) \tag{4.22}$$

The weights are updates according to the rule

$$\omega \leftarrow (1 - \epsilon\alpha)\omega - \epsilon\nabla_{\omega}J(\omega;y_{true}) \tag{4.23}$$

We can see that the weight decay term is now multiplicatively shrinking the weight vector by a constant factor in each step, before performing the usual gradient update.

4.6.1.2   Regularisation L1

Here the regularisation term is defined as

$$\Omega(\omega) = \|\omega\|_1 = \sum_i |\omega_i| \tag{4.24}$$

which leads us to have the objective function as

$$\widetilde{J}(\omega;y_{true}) = \alpha\|\omega\|_1 + J(\omega;y_{true}) \tag{4.25}$$

and a corresponding gradient

$$\nabla_{\omega}\widetilde{J}(\omega;y_{true}) = \alpha\,\mathrm{sign}(\omega) + \nabla_{\omega}J(\omega;y_{true}). \tag{4.26}$$

By observing the gradient we can notice how the gradient is scaled by a constant factor with a sign equal to $\mathrm{sign}(\omega_i)$.

4.7   Conclusions

In conclusion, the theory and mathematics behind ANNs have been addressed in this chapter. Fundamental notions such as activation functions, back propagation, gradient descent, overfitting and underfitting have been discussed. The next chapter will introduce the convolutional neural networks theory.

References

[1]   Géron A. *Hands-on Machine Learning with Scikit-Learn, Keras, and TensorFlow: concepts, Tools, and Techniques to Build Intelligent Systems.* O'Reilly Media; 2019.
[2]   Goodfellow I., Bengio Y., Courville A. *Deep Learning.* MIT press; 2016.
[3]   Vemuri V.K. *The Hundred-Page Machine Learning Book. Andriy Burkov, Quebec City, Canada, 2019, 160 pp., 49.99 (Hardcover); 29.00 (paperback); 25.43 (KindleEdition),(Alternatively, can purchase at leanpub.com at a minimum price of 20.00), ISBN 978-1999579517.* Taylor & Francis; 2020.
[4]   Jain A.K., Jianchang M., Mohiuddin K.M. 'Artificial neural networks: a tutorial'. *Computer.* 1996;**29**(3):31–44.

[5] LeCun Y., Bengio Y., Hinton G. 'Deep learning'. *Nature*. 2015;**521**(7553):436–44.

[6] Ruder S. 'An overview of gradient descent optimization algorithms'. arXiv preprint arXiv; 2016.

[7] Hochreiter S., Younger A.S., Conwell P.R. 'Learning to learn using gradient descent'. *International Conference on Artificial Neural Networks*. Springer; 2001. pp. 87–94.

[8] Si A. 'Backpropagation and stochastic gradient descent method'. *Neurocomputing*. 1993;**5**(4-5):185–96.

[9] Du S., Lee J., Li H., *et al*. 'Gradient descent finds global minima of deep neural networks'. *International Conference on Machine Learning*; 2019. pp. 1675–85.

[10] Bottou L. 'Large-scale machine learning with stochastic gradient descent'. *Proceedings of COMPSTAT'2010*. Springer; 2010. pp. 177–86.

[11] Zinkevich M., Weimer M., Li L., *et al*. 'Parallelized stochastic gradient descent'. *Advances in Neural Information Processing Systems*; 2010. pp. 2595–603.

[12] Bottou L. 'Stochastic gradient descent tricks'. *Neural Networks: Tricks of the Trade*. Springer; 2012. pp. 421–36.

[13] Zhang T. 'Solving large scale linear prediction problems using stochastic gradient descent algorithms'. *Proceedings of The Twenty-First International Conference on Machine learning*; 2004. p. 116.

[14] Hinton G., Srivastava N., Swersky K. 'Neural networks for machine learning lecture 6A overview of mini-batch gradient descent.'. *Cited on*. 2012;**14**(8).

[15] Keskar N.S., Mudigere D., Nocedal J. 'On large-batch training for deep learning: generalization gap and sharp minima'. arXiv preprint arXiv; 2016.

[16] Tan H.H., Lim K.H. 'Vanishing gradient mitigation with deep learning neural network optimization'. *2019 7th International Conference on Smart Computing & Communications (ICSCC)*. IEEE; 2019. pp. 1–4.

[17] Hanin B. 'Which neural net architectures give rise to exploding and vanishing gradients?' *Advances in Neural Information Processing Systems*; 2018. pp. 582–91.

[18] Hochreiter S. 'The vanishing gradient problem during learning recurrent neural nets and problem solutions'. *International Journal of Uncertainty, Fuzziness and Knowledge-Based Systems*. 1998;**06**(02):107–16.

[19] He K., Zhang X., Ren S., *et al*. 'Deep residual learning for image recognition'. *Proceedings of the IEEE conference on computer vision and pattern recognition*. 2015:770–8.

[20] Yang GSchoenholz S.S2018Deep mean field theory: Layerwise variance and width variation as methods to control gradient explosionhttps://openreview. net/forum?id=rJGY8GbR-

[21] Sun R. 'Optimization for deep learning: theory and algorithms'. arXiv preprint arXiv; 2019.

[22]   Weinan E., Yu B. 'The deep Ritz method: a deep learning-based numerical algorithm for solving variational problems'. *Communications in Mathematics and Statistics*. 2018;**6**(1):1–12.

Chapter 5

Convolutional neural networks

Silvia Liberata Ullo[1], Alessandro Sebastianelli[1], Maria Pia Del Rosso[1], Dario Spiller[2], Erika Puglisi[3], Artur Nowakowski[3], Mario Luca Bernardi[1], and Marta Cimitile[4]

In this chapter we introduce the convolutional neural network theory including concepts such as convolution operator, kernel, stride, padding and pooling.

5.1 Introduction

A convolutional neural network (CNN, or ConvNet) is a class of deep neural networks for processing data with a grid-like topology [1]. It is most commonly applied to analysing time-series data and visual imagery [2].

CNNs were inspired by discoveries in neuroscience trying to ask a question on how a mammal brain is structured, especially the part of it responsible for processing visual information [3]. Starting from the early 1960s, when the first two major types of cells in the visual cortex were identified [4], research on a model of a visual cortex has been carried out and arrived at the first functional model called neocognitron in 1980 [5]. Since then many similar models have been developed, but CNNs started gaining real popularity since 2012 when the AlexNet-based network outperformed the state-of-the-art solutions on the ImageNet Large Scale Visual Information Challenge [6].

CNNs are regularised versions of multilayer perceptrons. Multilayer perceptrons usually refer to fully connected networks, where each neuron in one layer is connected to all neurons in the next layer. The 'fully-connectedness' of these networks makes them prone to overfitting data. Typical ways of regularisation include adding some form of magnitude measurement of weights to the loss function (Chapter 4). However, CNNs take a different approach towards regularisation: they take advantage of the hierarchical patterns in data (i.e. the information in the input data is

[1]Engineering Department, University of Sannio, Italy
[2]Joint ASI - Φ-Lab, Italian Space Agency (ASI) and European Space Agency (ESA), Italy
[3]Φ-Lab, European Space Agency (ESA), Italy
[4]Unitelma Sapienza University of Rome, Italy

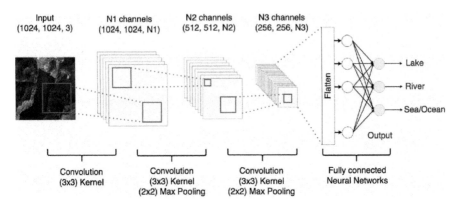

Figure 5.1   *An example of convolutional neural networks (CNNs) that classify Sentinel-2 images*

collected in well-defined features at different scales) and assemble more complex patterns using smaller and simpler structures in the network. Thinking about remote sensing images, for instance, there are usually some recurrent patterns (e.g. roads, building, green areas) both at a low resolution scale (e.g. clusters of building in a town, agglomerates of cultivated fields) and at a high resolution scale (e.g. a roof, a maize field). Such recurring patterns can be detected using the same analysis tool within the neural network, which happens to be the convolution operator related to optimised convolutional filters. Therefore, on the scale of connectedness and complexity, CNNs are at the lower extreme. They are also known as shift invariant or space invariant artificial neural networks., based on their shared-weights architecture and translation invariance characteristics.

CNNs use relatively little pre-processing compared with other image classification algorithms. This means that the network learns the filters that in traditional algorithms were hand-engineered. This independence from prior knowledge and human effort in feature design is a major advantage. This asset comes at a cost of more samples needed to train the network to avoid overfitting. CNNs have applications in image and video recognition, recommender systems, image classification, medical image analysis and natural language processing.

For example, as you can see in Figure 5.1, a CNN can classify Sentinel-2 images. In this easy example, the network takes as input a Sentinel-2 RGB (Red-Green-Blue) image and returns a prediction, that indicates if there is either a lake, a river or sea/ocean in the image. By leveraging this example, it is easy to understand that this kind of network can be applied to a huge variety of applications, even the more complex ones. In the following sections of this chapter we will explain how each block that constitutes a convolutional network works. As you can see from Figure 5.1, a standard classifying network is made up of a convolutional network and a fully connected neural network. The convolutional network has the responsibility of extracting and learning information from the data, also known as *feature extraction* phase, while the fully connected network deals with the classification of the inputs. The interested reader can find another example of CNN classifier applied to landslides in [7].

5.2   Convolutional neural networks

The name 'Convolutional Neural Network' derives from the mathematical operation used by this network: the convolution. A CNN consists, like a classic deep learning network, of an input and output layer as well as a series of hidden layers. Typically, the hidden layers perform the convolution operations based on convolutional filters usually referred to as *kernels*. Such kernels are associated with matrices (or n-dimensional tensors in generic cases) performing the convolution by means of well-known, simple mathematical rules. The activation function commonly used is the rectified linear unit function (ReLU) (see section 5.5), followed typically by additional layers of pooling (see section 5.6) and normalisation (see section 5.7).

5.2.1   *Motivation*

Convolution introduces three important ideas that can help improve a machine learning system: *sparse interactions*, *parameter sharing* and *equivariant representations*. Traditional neural network layers use matrix multiplication involving a matrix of neuron parameters and input data to describe the interaction between each input and output unit (Chapter 4). This means that every output unit interacts with every input unit. On the contrary, convolutional networks are mainly characterised by sparse interactions.

This can be done by making the kernel smaller than the input. For example, when processing an image, the input image can have thousands or millions of pixels, but small meaningful features (e.g. edges) can be detected with kernels that occupy only tens or hundreds of pixels. This allows to store fewer parameters, which reduces the memory requirements of the model, improves its statistical efficiency and allows to calculate the output with fewer operations. These improvements in efficiency are usually quite large. If there are m inputs and n outputs, then the matrix multiplication requires $m \times n$ parameters and the algorithms used in practice have $O(m \times n)$ runtime. If the number of connections for each output are limited to k, then the sparsely connected approach requires only $k \times n$ parameters and $O(k \times n)$ runtime. For many real applications, it is possible to obtain good performances while keeping k several orders of magnitude smaller than m [2].

For graphical demonstrations of sparse connectivity, see Figure 5.2. In a deep convolutional network, units in the deeper layers may indirectly interact with a larger portion of the input, as shown in Figure 5.3. This allows the network to efficiently describe complicated interactions between many variables by constructing such interactions from simple building blocks related to sparse interactions [2].

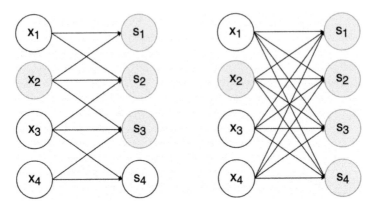

Figure 5.2   Sparse connectivity, where the input x_2 is highlighted as well as the output units s that are affected by this unit. (Left) When the s layer is formed by convolution with a kernel of width 3, only three outputs are affected by x. (Right) When the s layer is formed by matrix multiplication, connectivity is no longer sparse, so all of the outputs are affected by x_2

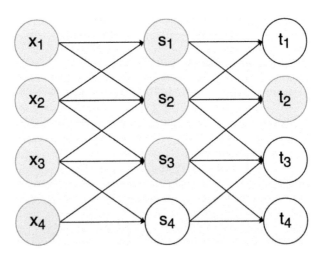

Figure 5.3   The receptive field of the units in the deeper layers of a convolutional network is larger than the receptive field of the units in the shallow layers. This effect increases if the network includes architectural features like strided convolution or pooling. This means that even though direct connections in a convolutional net are very sparse, units in the deeper layers can be indirectly connected to all or most of the input image

5.2.2 The convolution operation

In its most general form, convolution is an operation on two functions of a real-valued argument. To get an idea of what is convolution and how it works, we can start with an example. Suppose we are tracking the location of an object with a sensor. The sensor provides a single output $x(t)$, the position of the object at time t. Both x and t are real values. Let's suppose that the sensor measurement is affected by noise. To obtain a less noisy estimate of the object position, we can average together several measurements. Of course, more recent measurements are more relevant, so to take into account this fact a weighted average should be used. Let's define a weighting function $w(a)$, where a is the age of a measurement. By applying the weighted average operation at every moment, we obtain a new function $p(t)$ providing a smoothed estimation of the object position as

$$p(t) = \int x(a)w(t-a)da. \qquad (5.1)$$

This operation is called *convolution*. The convolution operation is also denoted with the asterisk notation $*$ as expressed in (5.2)

$$p(t) = (x * w)(t). \qquad (5.2)$$

In this example, w must be a valid probability density function, or the output is not a weighted average. Also, w has to be 0 for all negative arguments, or it will look into the future. These limitations are related to the particular example. In general, convolution is defined for any functions for which the above integral is defined, and may be used for other purposes besides taking weighted averages. In mathematics, in particular in functional analysis, convolution is a function between two functions of a variable which consists of integrating the product between the first and second translations of a certain value (see Figure 5.4).

Convolution is used in various fields of physics, statistics, electronics, image analysis and computer graphics.

In CNN terminology, the first argument of the convolution is often referred to the input, the second argument to the kernel and the output is sometimes referred to the feature map. More details are discussed in Section 5.4.

In the above example, we have considered a sensor able to provide a measurement at every instant of time but is not realistic. In fact, when we work with data on a computer, time will be discretised. Let's assume that the sensor can provide a new acquisition each second, the time index t can assume only integer values and so the discrete convolution can be defined as

$$p(t) = (x * w)(t) = \sum_{a=-\infty}^{\infty} x(a)w(t-a) \qquad (5.3)$$

In machine learning applications, the input is usually a multidimensional array of data and the kernel is usually a multidimensional array of parameters that are

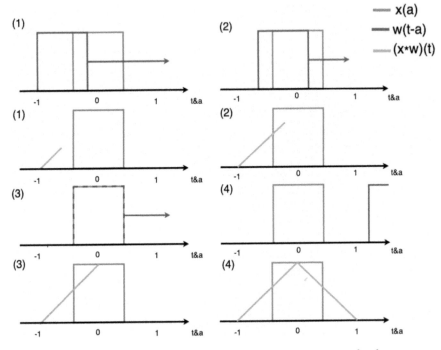

Figure 5.4   Example of convolution between two squared pulses

adapted during the learning/training process. These multidimensional arrays are usually referred to as tensors[a].

In deep learning, convolutions often apply over more than one axis simultaneously. For example, for a two-dimensional input image I, a two-dimensional kernel K should be used. The two-dimensional convolution formula applied to a 2D image is

$$S(i,j) = (I * K)(i,j) = \sum_m \sum_n I(m,n)K(i-m,j-n). \tag{5.4}$$

Convolution is a commutative operator, so (5.4) can also be written as

$$S(i,j) = (K * I)(i,j) = \sum_m \sum_n I(i-m,j-n)K(m,n). \tag{5.5}$$

Many neural network libraries implement a related function called the cross-correlation, which is the same as convolution but without flipping the kernel. Since this is only a trick for computational efficiency and it does not affect the results,

---

[a]In mathematics, a tensor is an algebraic object that describes the relationship between sets of algebraic objects related to a vector space. Objects that tensors may map include vectors and scalars, and, recursively, even other tensors. Tensors can take several different forms – for example: scalars and vectors (which are the simplest tensors), dual vectors, multilinear maps between vector spaces, and even some operations such as the dot product. Tensors are defined independently of any basis, although they are often referred to by their components on a basis related to a particular coordinate system. However, the AI community defines tensors as multidimensional arrays.

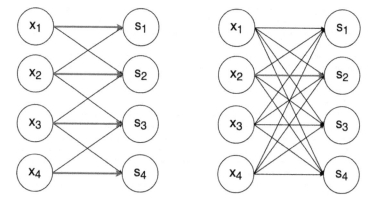

Figure 5.5   Parameter sharing: Red arrows indicate the connections that use
a particular parameter in two different models. (Left) The red
arrows indicate uses of the central element of a three-element kernel
in a convolutional model. Due to parameter sharing, this single
parameter is used at all input locations. (Right) The red arrow
indicates the use of the second element of the weight matrix in a
fully connected model. This model has no parameter sharing so the
parameter is used only once

cross-correlation is usually confused with convolution [2]. The cross-correlation is
described as

$$S(i,j) = (I * K)(i,j) = \sum_{m} \sum_{n} I(i + m, j + n)K(m, n). \tag{5.6}$$

5.2.3   Parameter sharing and sparse connectivity

Parameter sharing refers to using the same parameter for more than one function in
a model. In a traditional NN (Neural Network), each element of the weight matrix
is used once when computing the output of a layer. It is multiplied by one element
of the input and then never revisited. As a synonym for parameter sharing, one can
say that a network has tied weights, because the value of the weight applied to one
input is tied to the value of a weight applied elsewhere. In a CNN, each member of
the kernel is used at every position of the input. The parameter sharing used by the
convolution operation means that rather than learning a separate set of parameters
for every location, we learn only one set. This does not affect the runtime of forward
propagation (it is still $O(k \times n)$) but it does further reduce the storage requirements
of the model to k parameters. Recall that k is usually several orders of magnitude less
than m. Since m and n are usually roughly the same size, k is practically insignificant
compared to $m \times n$. Convolution is more efficient than dense matrix multiplication
in terms of the memory requirements and statistical efficiency [2]. For a graphical
description of how parameter sharing works, see Figure 5.5.

A common example of how both of these first two principles work is shown
in Figure 5.6, which demonstrates how sparse connectivity and parameter sharing

 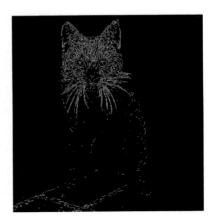

Figure 5.6   Efficiency of edge detection

can improve the efficiency for detection of edges in an image. The image on the right was formed by taking each pixel in the original image (on the left) and subtracting the value of its neighbouring pixel on the left. This shows the strength of all of the vertically oriented edges in the input image, which can be a useful operation for object detection. Both images have a size of 640 × 640 pixels. The edge detection can be described by a convolution kernel containing two elements, and requires 640 × 640 × 3 = 1.228.800 floating point operations (two multiplications and one addition per output pixel) to compute using convolution. To describe the same transformation with a matrix multiplication would take 640 × 640 × 640 × 640 = 167.772.160.000 entries in the matrix, making convolution more efficient for representing this transformation.

In the case of convolution, the particular form of parameter sharing causes the layer to have a property called *equivariance to translation*. A function is *equivariant* when the output changes in accordance to how the input changes. Specifically, a function $f(x)$ is equivariant to a function g if $f(g(x)) = g(f(x))$. In the case of convolution, if we let g be any function that translates the input, i.e., shifts it, then the convolution function is equivariant to g.

For example, let I be a function giving image brightness at integer coordinates. Let g be a function mapping one image function to another image function, such that $I' = g(I)$ is the image function with $I'(x,y) = I(x-1,y)$. This shifts every pixel of I one unit to the right. If we apply this transformation to I, then applying convolution, the result will be the same as if we applied convolution to I', then applied the transformation g to the output.

When processing time series data, this means that convolution produces a sort of time line that shows when different features appear in the input. If we move an event later in time in the input, its exact same representation will appear in the output, just later in time. Similarly with images, convolution creates a two-dimensional map of where certain features appear in the input. If we move the object in the input, its representation will move the same amount in the output. For example, when processing images, it is useful to detect edges in the first layer of a convolutional network. The same edges appear more or less everywhere in the image, so it is practical to share parameters across the entire image [2].

5.3   Networks and layers structure

As of today, there are plenty of different CNN structures in literature. The most interesting ones are those related to the ImageNet competition, as sometimes they really have revolutionised the world of machine learning. Among the most well-known CNN structures, it is worthy to mention Alexnet [8] (widely regarded as one of the most influential contributions in the field), VGG net [9] (VGG derives from the research group 'Visual Geometry Group'), GoogLeNet with its inception module [10], Microsoft ResNet [11] and region-based CNNs [12]. All these architectures are characterised by many convolutional layers and sometimes they have specialised structures, as the inception module in GoogLeNet or the residual connection in ResNet. However, there are some general principles that characterise the convolutional layers in all these generic architectures. Indeed, a typical layer of a CNN consists of four stages (see Figure 5.7):

- The convolution stage, where the layer performs several convolutions in parallel to produce a set of linear activations
- The detection stage, where each linear activation is run through a non-linear activation function, such as a ReLU
- The pooling stage, where a pooling function is used to modify the output of the layer further. A pooling function replaces the output at a certain location with a summary statistic of the nearby outputs
- The normalisation stage, usually performed as batch normalisation. To increase the stability of a neural network, batch normalisation normalises the output of

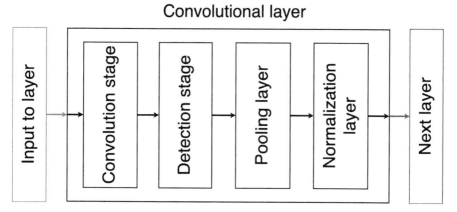

Figure 5.7   The components of a typical convolutional neural network layer

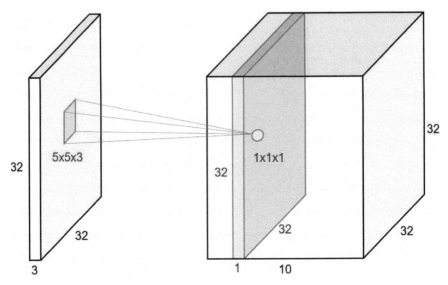

Figure 5.8   *Example of how the convolution operation is performed on an*
image. The matrix multiplication between the filter (the 5×5×3
square on the left) and a part of the input image forms an element
(the circle on the right) of the feature map (represented in dark blue
on the right). In this image ten different filters are used, this is why
the output image has a size of 32 × 32 × 10

a previous activation layer by subtracting the batch mean and dividing by the
batch standard deviation.

5.4   Kernels and feature maps

The main building block of a CNN is the convolutional stage in which the convolution operation is applied on the input data using a kernel/filter to produce a feature
map. Each layer contains a series of filters, which are matrices of integers that slide
over the input image according to stride and padding values (see section 5.4.1 and
section 5.4.3 for further details). At every location, an element-wise matrix multiplication is performed, and the result is summed and goes into the feature map. Usually
multiple convolutions are performed on an input, each using a different filter and
resulting in a distinct feature map. All these feature maps are then stacked together
to assemble the final output of the convolution layer (see Figure 5.8).

Typically hundreds of kernels or filters are used in a CNN, so the network is
able to extract different information/features. For example, in Figure 5.9 the first
three filters of size 3 × 3 of the VGG16 network are represented [13].

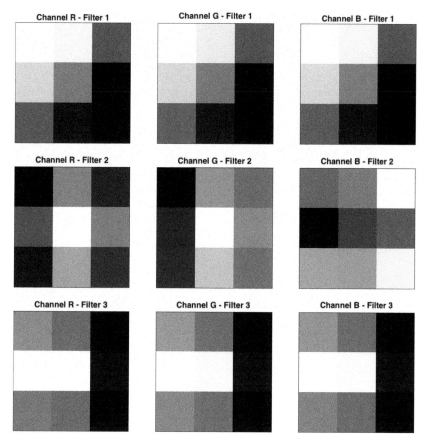

*Figure 5.9   First three filters of the first layer of the VGG16 network [13], one
row for each filter and one column for each channel*

The feature maps, also called activation maps, are the result of each convolution
operation between an input image and a kernel, and capture the result of applying
the filters to an input (which can be the input image or another feature map).

Usually, the first layer feature maps retain most of the information present in
the image. In CNN architectures the first layers typically act as *edge detectors*;
while moving deeper into the network the feature maps look less like the original
image and more like an *abstract representation* of it (see Figure 5.10). The reason
is that deeper feature maps encode high-level concepts like 'cat nose' or 'dog ear'
while lower-level feature maps detect simple edges and shapes. That's why deeper
feature maps contain less information about the image and more about the class
of the image. Most of the times, they still encode useful features, but they are
less visually interpretable by humans. The feature maps become sparser as we go
deeper, meaning the filters detect less features. It makes sense because the filters in

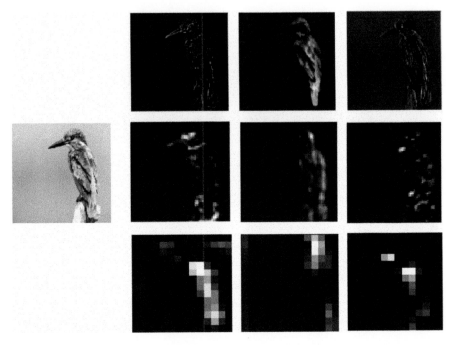

Figure 5.10   The original image (in RGB on the left) and the feature maps generated by the filters of the VGG16 network (on the right). In the first row it is possible to see three feature maps generated by the first layer of the network (in particular the three ones generated by the three filters represented in Figure 5.9); in the second row the feature maps generated by the ninth layer and in the last row the feature maps generated by the sixteenth layer

the first layers detect simple shapes, and every image contains those. But as we go deeper we start looking for more complex stuff like 'dog tail' and they don't appear in every image.

5.4.1   Stride

The kernel shifts over the input with rules defined by the stride value, every time performing a matrix multiplication operation between the kernel K and the portion P of the input over which the kernel is hovering. The movement of the kernel is showed in Figure 5.11.

The kernel moves from the top left to the bottom right of the input with a certain stride value that specifies how many columns/rows the kernel has to slide while moving to the right/down. An example of convolution operation with stride length equal to 1 is shown in Figure 5.12.

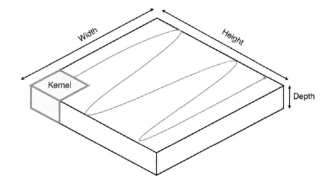

Figure 5.11   Movement of the kernel represented by the blue line

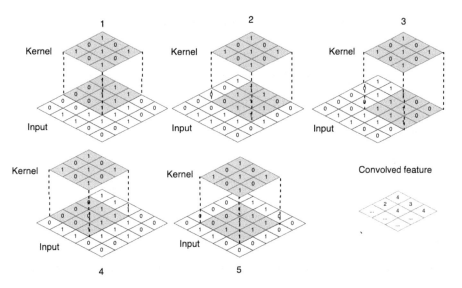

Figure 5.12   Convolution operation with stride length = 1

5.4.2   Multidimensional input

In the case of multidimensional input (e.g. RGB images), the kernel has the same depth as that of the input. Denoting with 1, 2, 3 the red, green and blue channels, respectively, matrix multiplication is performed between K and I stack ($[K_1, I_1]; [K_2, I_2]; [K_3, I_3]$) and all the results are summed with the bias. The final output is a one-depth convoluted feature; an example is shown in Figure 5.13.

5.4.3   Padding

During the convolution operation, the position of the filter window is updated according to the stride argument, but when the border pixels are reached some of

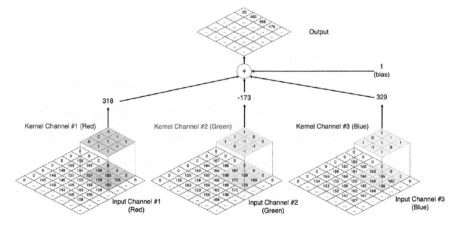

Figure 5.13   *Convolution operation on an M × N × 3 image matrix with a*
3 × 3 × 3 kernel

the elements of the filter may not overlap the input elements. Therefore, in order to compute the values of those border regions, it is possible to 'extend' the image applying the *same padding* rule. However, if these border regions need to be discarded, then a *valid padding* strategy is applied, meaning no padding is applied at all. It is important to specify that the rows and columns of the padding are always filled with zero values.

As can be seen from Figure 5.14, the *same* padding value of 1 is applied to the original 4 × 4 × 1 image, so its dimension becomes 6 × 6 × 1 because additional rows and columns are added at the top, bottom, left and right of the input matrix. At this moment, if a 3 × 3 × 1 kernel is applied to the image then the convoluted matrix will be of dimensions 4 × 4 × 1, exactly like the original input image. This means that the use of the padding preserves the information contained in the original image. On the other hand, if the same operation is performed without the use of the padding, meaning a *valid* padding is used, then the output will have a dimension of 2 × 2 × 1.

5.5 ReLU layer

The rectified linear unit is the most commonly used activation function in deep learning models. As explained in Chapter 4, the function returns zero if it receives any negative input, but for any positive value x it returns that value back. The mathematical formulation of this function is

$$f(x) = \max(0, x). \tag{5.7}$$

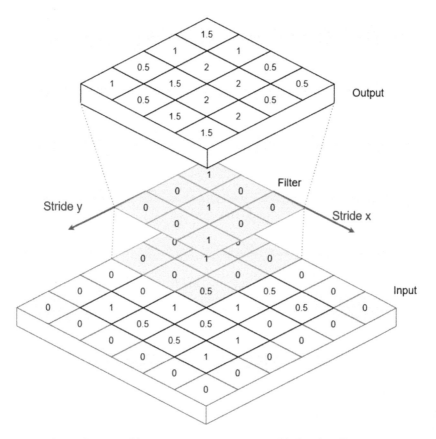

Figure 5.14   *Same padding: 6 × 6 × n image is padded with a filter to create a 4 × 4 × n image*

It is interesting to underline that this simple function can allow the model to account for non-linearities and interactions. For these reasons this function works great in most applications, and it is very widely used as a result.

Activation functions have two primary purposes:

- Help a model to find out interaction effects: an interaction effect is when a variable affects a prediction depending on the value of associated variables
- Help a model to introduce non-linear effects: without non-linearity in the network the output will just be a linear transformation of the input, which is not powerful enough to approximate non-linear real world problems.

ReLU captures interactions and non-linearities with the following mechanisms:

1. **Interactions:** Imagine a single node in a neural network model. For simplicity, assume it has two inputs, called *A* and *B*. The weights from *A* and *B* into our

node are 2 and 3, respectively. So the node output is $f(2A + 3B)$. Applying the ReLU function to f, we get that if $2A + 3B$ is positive, the output value of our node is also $2A + 3B$ and if $2A + 3B$ is negative, the output value of our node is 0. Let us consider a case where $A = 1$ and $B = 1$. The output is $2 + 3 = 5$, and if A increases, then the output increases too. If $B = -100$ then the output is 0, and if A increases moderately, the output remains 0. So A might or might not increase the output depending on the value of B. This is a simple case where the node captured an interaction. With more nodes and more layers, the power and complexity of interactions increases.

2. **Non-linearities:** A function is non-linear if the slope is not constant. The ReLU function is non-linear around 0, but the slope is always either 0, for negative values, or 1, for positive values. There are two ways in which the combination of ReLU nodes leads to the capture of different non-linearities. First, most models include a bias term for each node. The bias term is just a constant number that is determined during model training (Chapter 4). For simplicity, consider a node with a single input called A, and a bias. So the bias term allows us to move where the slope changes. So far, it still appears we can have only two different slopes. However, real models have many nodes. Each node (even within a single layer) can have a different value for it's bias, so each node can change slope at different values for our input. When we add the resulting functions back up, we get a combined function that changes slopes in many places. These models have the flexibility to produce non-linear functions and account for interactions well. As we add more nodes in each layer the model gets even greater ability to represent these interactions and non-linearities.

5.6   Pooling

A limitation of the feature map output of convolutional layers is that they record the precise position of features in the input: this means all the movements in the position of the feature (e.g. rotations, shifts, etc.) will result in a different feature map. To address this problem usually a *downsampling* of the input signal is needed, so, a lower resolution version of the input that still contains the important structural elements (without the fine details) is created. The downsampling operation can be performed changing the stride of the convolution across the image or using a pooling layer. The pooling layer helps to make the representation become approximately invariant to small translations of the input and, therefore, to improve the statistical efficiency of the network. In other words, it reduces the spatial size of the convolved feature to decrease the computational power required to process the data.

There are different kinds of pooling (see Figure 5.15):

•   Max pooling: It returns the maximum pixel value from the portion of image, called *patch*, covered by the kernel. Usually this pooling works better in

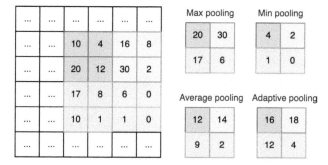

Figure 5.15   *A visual presentation of how different kinds of pooling downsample the pixel values*

computer vision tasks like image classification, because it highlights the most present feature in the patch

- Min pooling: It returns the minimum pixel value from the patch covered by the kernel
- Average pooling: It returns the rounded up average of all the pixel values from the patch covered by the kernel
- Adaptive pooling: Different to other kinds of pooling, in this one it is necessary to specify the output size and, then, the stride and the kernel size are automatically selected to match the needs. This kind of pooling is implemented in some libraries, like PyTorch
- Global pooling: It downsamples not the patch covered by the kernel, but the entire feature map to a single value. It can be used to aggressively summarise the presence of a feature in an image or as an alternative to a fully connected layer.

5.7   Batch normalisation

Batch normalisation is a technique for improving the speed, performance and stability of artificial neural networks. It is used to normalise the input layer by re-centring and re-scaling [14–16]. The effectiveness of the batch normalisation is evident, but the reasons behind that are currently under discussion and sometimes the reasons are discordant. Interested readers can refer to Knyazev and Neymeyr, Kohler *et al.*, Yang *et al.* and Santurkar *et al.* [17–20], for a detailed study on the effectiveness of batch normalisation.

5.7.1   Motivation

Normalisation is a common and useful technique when we work with the machine learning algorithm. For example, when we have features from 0 to 1 and some from 1 to 300, we should normalise them to speed up learning. Since the input layer is benefiting by it, the same procedure can also be followed for the values in the hidden layers, that are changing all the time, to get 10 times or more improvement in the training speed.

Batch normalisation reduces the variability of hidden unit values shift (covariance shift). If an algorithm learnt some X to Y mapping, and if the distribution of X changes, then we might need to retrain the learning algorithm by trying to align the distribution of X with the distribution of Y.

Batch normalisation allows each layer of a network to learn more by itself, independently of the other layers.

- We can use higher learning rates because batch normalisation makes sure that there is no activation that has gone really high or really low.
- It reduces overfitting because it has a slight regularisation effect. Similar to dropout, it adds some noise to each hidden layer's activations. Therefore, if we use batch normalisation, we will use less dropout, which is a good thing because we are not going to lose a lot of information. However, we should not depend only on batch normalisation for regularisation; we should better use it together with dropout.

5.7.2   How does batch normalisation work?

To increase the stability of a neural network, batch normalisation normalises the output of a previous activation layer by subtracting the batch mean and dividing by the batch standard deviation. However, after this shift of activation outputs , the weights in the next layer are no longer optimal. Stochastic gradient descent (SGD) undoes this normalisation if it is a way for it to minimise the loss function. Consequently, batch normalisation adds two trainable parameters to each layer, so the normalised output is multiplied by a 'standard deviation' parameter (γ) and adds a 'mean' parameter (β). In other words, batch normalisation lets SGD do the denormalisation by changing only these two weights for each activation, instead of losing the stability of the network by changing all the weights.

5.8   Common tasks performed by CNNs

Today CNNs are used for several different tasks, and some of them will be described in the following chapters of this book. Among the most exploited application, it is worthy to mention that CNNs can be used for *classification, localisation, detection* and *segmentation*.

The classification task corresponds to the process of taking an input image and outputting a class number out of a set of categories. In this case, the last layers of the networks are fully connected layers.

Differently, when we take a task like object localisation, our job is to produce not only a class label but also a bounding box that describes where the object is in the picture. Also in this case, the last layers of the network are fully connected, even though the structure is slightly more complicated with respect to the classification case. When dealing with detection, localisation needs to be done on all of the objects in the image. Therefore, you will have multiple bounding boxes and multiple class labels.

Finally, in the case of image segmentation, the task is to create a mask image where each pixel is associated to a metadata information describing the content of the image. In this case, we usually talk about fully convolutional networks, and one of the most famous architectures is the U-Net [21].

To conclude, it is noteworthy to mention one technique that is progressively gaining more and more interest in the field of deep learning, that is, the concept of *transfer learning* [22]. A common misconception in the deep learning community is that without a *Google-esque* amount of data, you can't possibly hope to create effective deep learning models. While data is a critical part of creating the network, the idea of transfer learning has helped to lessen the data demands. Transfer learning is the process of taking a pre-trained model (the weights and parameters of a network that has been trained on a large dataset by somebody else) and 'fine-tuning' the model with your own (small) dataset. The idea is that this pre-trained model will act as a feature extractor. The last layer of the network is usually removed and replaced with your own classifier (depending on what your problem space is). You then freeze the weights of all or some of the other layers and train the network normally (freezing the layers means not changing the weights during learning).

5.9   Conclusions

In conclusion, the theory and mathematics behind CNNs have been explained in this chapter. Fundamental notions like the convolution operator, kernal and features maps, and batch normalisation have been discussed.

The next chapter will introduce the reader to pre-processing and post-processing techniques for downloading and elaborating remote sensing datasets for AI (Artificial Intelligence) applications.

References

[1]   LeCun Y., Kavukcuoglu K., Farabet C. 'Convolutional networks and applications in vision'. *Proceedings of 2010 IEEE International Symposium on Circuits and Systems*. IEEE; 2010. pp. 253–6.

[2]   Goodfellow I., Bengio Y., Courville A. *Deep Learning*. The MIT Press; 2016.

[3]   Lindsay G.W. 'Convolutional neural networks as a model of the visual system: Past, present, and future'. *Journal of Cognitive Neuroscience*. 2020:1–15.

[4]   Hubel D.H., Wiesel T.N. 'Receptive fields, binocular interaction and functional architecture in the cat's visual cortex'. *The Journal of Physiology*. 1962;**160**(1):106–54.

[5]   Fukushima K., Miyake S. 'Neocognitron: a self-organizing neural network model for a mechanism of visual pattern recognition'. *Competition and Cooperation in Neural Nets*. Springer; 1982. pp. 267–85.

[6]   Krizhevsky A., Sutskever I., Hinton G.E. 'Imagenet classification with deep convolutional neural networks'. *Advances in Neural Information Processing Systems*; 2012. pp. 1097–105.

[7]   Ullo S.L., Langenkamp M.S., Oikarinen T.P., *et al*. 'Landslide geohazard assessment with convolutional neural networks using sentinel-2 imagery data'. *IGARSS 2019-2019 IEEE International Geoscience and Remote Sensing Symposium*. IEEE; 2019. pp. 9646–9.

[8]   Krizhevsky A., Sutskever I., Hinton G.E. 'ImageNet classification with deep convolutional neural networks'. *Communications of the ACM*. 2017;**60**(6):84–90.

[9]   Simonyan K., Zisserman A. 'Very deep convolutional networks for large-scale image recognition'. arXiv preprint arXiv; 2014.

[10]   Szegedy C., Liu W., Jia Y., *et al*. 'Going deeper with convolutions'. *Proceedings of the IEEE Conference on Computer Vision and Pattern Recognition*; 2015. pp. 1–9.

[11]   He K., Zhang X., Ren S., *et al*. 'Deep residual learning for image recognition'. *Proceedings of the IEEE conference on computer vision and pattern recognition*. 2015:770–8.

[12]   Girshick R., Donahue J., Darrell T., *et al*. 'Rich feature hierarchies for accurate object detection and semantic segmentation'. *Proceedings of the IEEE Conference on Computer Vision and Pattern Recognition*; 2014. pp. 580–7.

[13]   Simonyan K., Zisserman A. 'Very deep convolutional networks for large-scale image recognition'. arXiv preprint arXiv; 2014.

[14]   Ioffe S., Szegedy C. 'Batch normalization: accelerating deep network training by reducing internal covariate shift'. arXiv preprint arXiv; 2015.

[15]   *Glossary of Deep Learning: Batch Normalisation [online]*. Medium; 2017. Available from https://medium.com/deeper-learning/glossary-of-deep-learning-batch-normalisation-8266dcd2fa82 [Accessed 27 Jul 2017].

[16]   Batch normalization in Neural Networks [online]. Towards Data Science. 2017. Available from https://towardsdatascience.com/batch-normalization-in-neural-networks-1ac91516821c.

[17]   Knyazev A.V., Neymeyr K. 'A geometric theory for preconditioned inverse iteration III: a short and sharp convergence estimate for generalized eigenvalue problems'. *Linear Algebra and its Applications*. 2003;**358**(1):95–114.

[18]   Kohler J., Daneshmand H., Lucchi A., *et al*. 'Exponential convergence rates for batch normalization: the power of length-direction decoupling in non-convex optimization'. *The 22nd International Conference on Artificial Intelligence and Statistics*; 2019. pp. 806–15.

[19]   Yang G., Pennington J., Rao V. A mean field theory of batch normalization. arXiv preprint arXiv; 2019.

[20]   Santurkar S., Tsipras D., Ilyas A., *et al*. 'How does batch normalization help optimization?' *Advances in Neural Information Processing Systems*; 2018. pp. 2483–93.

[21]   Ronneberger O., Fischer P. 'Brox T. U-Net: convolutional networks for bio-medical image segmentation'. *International Conference on Medical image computing and computer-assisted intervention. Springer.* 2015:234–41.

[22]   Yosinski J., Clune J., Bengio Y. 'How transferable are features in deep neural networks?' *arXiv preprint.* 2014arXiv:1411.1792.

Chapter 6

How to create a proper EO dataset

Maria Pia Del Rosso[1], Alessandro Sebastianelli[1], Sara Aparicio[2], and James Wheeler[3]

In this chapter we describe how to create and manage a good satellite imagery data-set. We first focus on a classic way to get data, and then explain how to create a script to get it automatically, showing the main steps through code snippets and pseudo code blocks. Finally, a description of how to annotate this data will be given, thus generating the EO (Earth Observation) training datasets. Authors will mainly focus on Sentinel data from the Copernicus programme.

6.1   Introduction

This chapter focuses on what is perhaps the most important element and the core of supervised machine learning: *the training dataset*. This is actually the input data that will train supervised AI (Artificial Intelligence) models and strongly determine the performances of any data-driven learning processes. Previous chapters have already addressed some of the key elements that contribute to a good performing ML (Machine Learning) algorithm. The success of an ML task relies heavily on the quality of the dataset – and in fact this is considered to be the bottleneck of AI, as training datasets, specially for earth observations, are scarce and their preparation is very time-consuming. In supervised learning, an EO training dataset includes pairs of satellite imagery (*input x*) and their labels (*input y*).

As mentioned in Chapter 3.3, a training dataset needs to be **representative**, in the sense it encompasses all of the features and phenomena that you would want to identify, map or monitor. For example, in a classification problem where the goal is to classify Sentinel-2 scenes depending on an urban feature (like 'airports', 'ports', 'random', 'stadium') the training dataset should include several examples for each of these classes. Another example could be the per-pixel classification of the land cover type within 10 classes (e.g. 'water', 'forest', 'grass', 'bare soil', etc.). In such

[1]Engineering Department, University of Sannio, Benevento, Italy
[2]Solenix/ESA, Rome, Italy
[3]Φ-Lab, European Space Agency (ESA), Frascati, Italy

a case, the training dataset should include examples of all the ten classes. In both examples, a major problem for real applications is represented by the equal distribution of samples per classes, as usually EO datasets are strongly imbalanced.

The **quality** of training data is crucial to achieve good results in machine learning algorithms as it avoids bringing in data with errors, outliers and noise which may cause the system to perform less well. In Chapter 2.6 classical workflows to pre-process Sentinel-1 and Sentinel-2 data are described.

The **quantity** of input data is also a key element that will dictate the success of an ML problem, even though there is no recipe to tell if the amount of data you have is 'enough' to allow the model to generalise well. Indeed, the correct amount of input data depends on several factors, such as the complexity of the problem or the complexity of the learning algorithm (e.g. deep learning models are often more 'data hungry'). However, the creation of training datasets is often a time-consuming process (specially if one considers the labelling/annotating task) which could often result in little datasets. Some approaches to overcome problems related to small input datasets, such as the data augmentation approach, have already been discussed in Chapter 3. A proper dimension of the input dataset is of crucial importance when an extensive training and validation campaign is required. Tasks like hyperparameter optimization or cross-validation requires the splitting of the input dataset into training, validation and test datasets, and such splitting only makes sense when the input dataset is large enough. Indeed, the distribution of the features and classes into training, validation and test datasets should be as similar as possible in order to carry out a consistent analysis.

Although seen as one of the greatest challenges in AI for earth observation applications, there is already a growing number of publicly available EO training datasets [1–4] in websites like Kaggle or OpenMl. However, if you are addressing a problem for which there is no available dataset repository, you will need to create your own dataset, and the following sections will introduce the reader to classical processes to retrieve data with labels.

6.2   Non-automated method

In this section a method to manually download and process satellite data will be presented. As stated before, this book will focus on Sentinel data within the Copernicus programme and consequently, in this section, a method for downloading Sentinel data will be described.

6.2.1   The open access hub

The Copernicus programme offers a platform to download Sentinel-1, Sentinel-2, Sentinel-3 and Sentinel-5P user products, called Open Access Hub. In order to access the platform [5], the user must be registered. The hub is shown in Figure 6.1.

To select the area of interest (AOI), the user should select first the button on the right (the square below the directional arrows) and then draw a polygon around the AOI, as shown in Figure 6.2.

Figure 6.1   The ESA open access hub – home page (source [5])

In order to select the Satellite platform, the type of data and the time interval, the user should first open the settings panel, by clicking on the menu on the top-left corner, and then set the parameters, as shown in Figure 6.3.

The user should then click on the save button and the search button. After clicking on the search button, the product list should appear as shown in Figure 6.4. By moving the mouse on the product list or on the map, a single product should be highlighted. Each product can be stored in the cart, can be directly downloaded and can be plotted for a quick look.

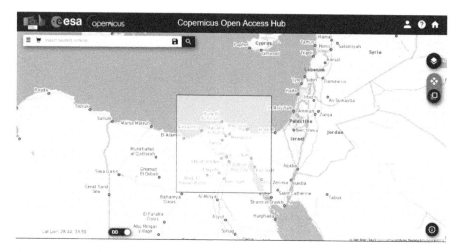

Figure 6.2   The ESA open access hub – select area of interest (AOI) (source [5])

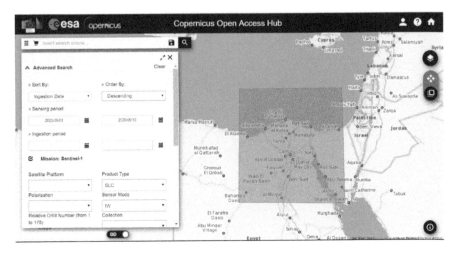

Figure 6.3   The ESA open access hub – select settings (source [5])

6.2.2   SNAP – Sentinel Application Platform

The Sentinel Application Platform (SNAP) is a common architecture for all Sentinel Toolboxes. The software is developed by Brockmann Consult, Skywatch, Sensar and C-S. SNAP and the individual Sentinel Toolboxes support numerous sensors other than Sentinel sensors. ESA is providing the SNAP user tool free of charge to the Earth Observation Community. Each toolbox consists of a collection of processing tools, data product readers and writers, and a display and analysis application to support the large archive of data.

SNAP can be downloaded from the official page [6], where the reader can also learn more about each toolbox and each functionality of the ESA software. In

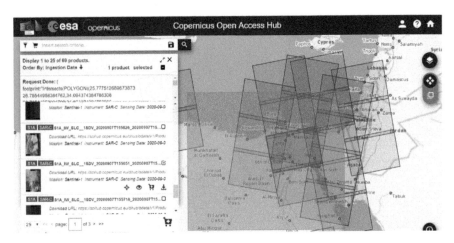

Figure 6.4   The ESA open access hub – download data (source[1])

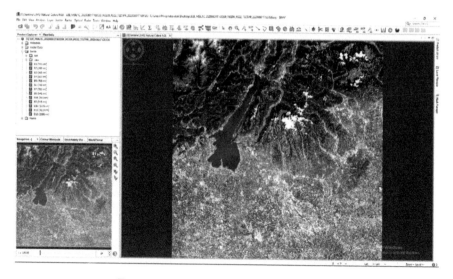

Figure 6.5   Sentinel Application Platform

Figure 6.5 the reader can see the GUI (Graphical User Interface) of SNAP, where a Sentinel-2 image is loaded.

SNAP allows the user to apply a huge variety of pre-processing techniques and to run several applications with Sentinels data. Interested readers can refer to [7] to learn the most important features of SNAP through guided tutorials. For the implementation of a dataset, SNAP can be used for the pre-processing of Sentinel products and then smaller images can be extracted from the product, using, for instance, Python scripts.

Since this procedure can be computationally heavy and more actions are needed to implement the final dataset, an automated method, presented in the next section, is used.

6.3   Automated method

In this section one of the most common tools for automatically downloading satellite data is presented: Google Earth Engine (GEE). This platform allows to analyse, process and download, not only Sentinel data, but also a large variety of satellite data. Google Earth Engine can be used online (and it is based on Javascript) or remotely using the Python API (Application Programming Interface). These functionalities will be explained in this section.

6.3.1   Google Earth Engine – online editor

The online editor allows the user to write JavaScript code to navigate and elaborate on data of the available catalogue. Each catalogue typically represents a satellite mission.

The online editor is composed of different panels (see Figure 6.6): the code editor (top centre), the script searcher (top left), the console (top right) and the map

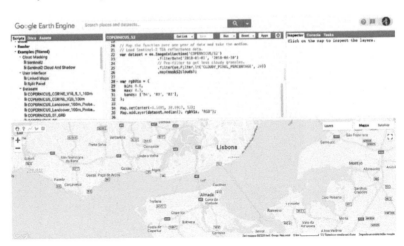

Figure 6.6   Google Earth Engine code editor

(bottom). The online editor also allows to display the satellite images on the map, to draw polygons or points of interest, to store user scripts, etc.

Google Earth Engine contains a lot of pre-defined scripts (these can be accessed through the script searcher) which can be used to learn the basic functionalities of the platform. For instance, results of the pre-defined Sentinel-2 example are shown in Figure 6.7.

In general the catalogue can be filtered using spatial filters, temporal filters or bands filters. These filters can be used to extract a specific image. The data can be downloaded, for example in a Google Drive folder, using the export function (see code block 6.6.1).

Since the platform contains an huge variety of functions and applications, interested users can refer to the official guide [8].

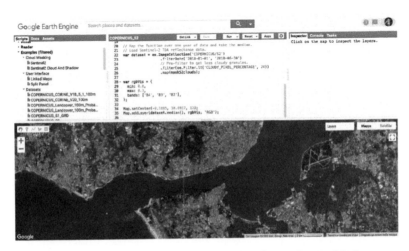

Figure 6.7   Google Earth Engine code editor – Sentinel-2 data

6.3.2   Google Earth Engine – Python API

Google Earth Engine can also be used through Python API. The authors suggest using Google Colab to work with the API. You can install the API in Colab, using *pip*, by launching the command *!pip install earthengine-api*. Before starting work with the API, the user must authenticate and initialise the python environment. By running the code in code block 6.3.2, a pop-up window appears in the browser, asking to login with the same google account used for the registration to GEE. After the login process, a new window shows a passcode to be copied in the Colab entry field.

Initialise Google Earth Engine

```
import ee
!earthengine authenticate
ee.Initilize()
```

Script for downloading Sentinel-1 GRD (Ground Range Detected) products

In this paragraph the authors want to show how to download Sentinel-1 data from Google Earth Engine, using Python API. To solve the goals some functions are necessary. First of all, as you can see in code block 6.3.2, the authors defined a function to select a specific image from the whole archive of Sentinel-1 products. The function *get_vv_data* uses as input a geometry, essentially a set of coordinates, and returns an image corresponding to these coordinates. The image is the median of all the products, acquired in VV polarisation, found in that location during a time interval of 1 year.

Get Data

```
# Apply a mask to the image using the angle of acquisition
def s1MaskBorder(image):
    angle = image.select('angle')
    mask = angle.gt(31).And(angle.lt(45))
    return image.updateMaks(mask)

def get_VV_data(geometry):
    # Load the Sentinel-1 Image Collection
    sentinel1 = ee.ImageCollection('COPERNICUS/S1_GRD_FLOAT')
    # Filter by metadata properties
    return (sentinel1
      .filter(
        ee.Filter.listContains('transmitterReceiverPolarisation', 'VV'))
      .filterBounds(geometry),
      .filterDate('2018-01-01', '2018-12-31'),
      .map(s1MaskBorder),
      .select('VV'),
      .median(),
      .clip(geometry))
```

The second function, shown in code block 6.3.2, is used to create an exporting task. The function uses the image generated by the previous function, the geometry and a name to create a Google Earth Engine task. This task is used to request the image from the GEE catalogue that will be saved in a Google Drive folder named *vv_dataset*. The image has a shape of 256×256

Export Data

```
def export(img, geom, name):
    task = ee.batch.Export.image.todrive(
        folder = 'vv_dataset',
        image = img,
        description = name,
        fileNamePrefix = name,
        region= geometry.coordinates().getInfo(),
        dimensions = "256x256",
        maxPixels = 1e9
    )

    task.start()
    print(task.status())
    return task
```

In order to download several images, the authors propose a simple loop, shown in code block 6.3.2, that generates random coordinates and downloads images using the functions previously described.

Download Data

```
from random import uniform

def create_buffer(x, y):
    return ee.Geometry.Point(x, y).buffer(1280).bounds()

number_of_images = 100
for i in range(number_of_images):
    # Random coordinates
    longitude = uniform(-180,180)
    latitude = uniform(-90, 90)

    geometry = create_buffer(longitude, latitude)
    s1_vv = get_VV_data(geometry)
    export(s1_vv, geometry, 'vv_' + str(i))
```

Obviously this is only a starting example, as mentioned before Google Earth Engine has an huge variety of data, so the interested reader can easily reorganise and modify the code to adapt it to different applications. A more advanced software based on the Google Earth Engine Python API is presented in [9], where a software with GUI has been implemented for non-expert users.

6.4   Dataset annotation

This chapter walked you through the accessing and 'grabbing' of the EO data that you will use for your machine learning process. However, one important step is still missing in case you want to perform supervised learning: the dataset needs to be *annotated* – or 'labelled'. This process is the core of supervised learning, as this is what will create the 'examples' that will teach the machine. The learning algorithms will learn the patterns through labelled data, which will allow it to identify the same patterns in *unseen* (i.e. unlabelled) data. However, different supervised learning tasks (e.g. image classification, image segmentation or object detection) will require different ways of annotating data in a format that the algorithm can use.

A word of caution: terminology

Not all training data will require to have 'labels' – but to be structured in a manner that tells the model what is the *target* to be predicted, we will see this example later as well. Training Dataset is the part of the Original Dataset that we use to train our ML model. The model learns on this data by running the algorithm and maps a function $F(x)$ where 'x' is the independent variable (inputs) for 'y', where 'y' is the dependent variable (output). While training our model, we provide both input and output variables so that our model is able to learn to predict the output based on the input data.

The following sections show different ways of creating training datasets – for three different tasks – image classification, semantic segmentation and object detection. There are currently several open source and paid annotation tools. For the sake of diversity, the authors give examples for each supervised learning task, using different popular, free and open annotation tools.

6.5 Training dataset for image classification tasks

As explained in Chapter 3, image classification involves assigning one (*or more*) class label to an entire image. Hence, the annotation of this data will consist of assigning a label to each single image. Image classification in the context of EO is often used to identify if the image contains a specific feature. This case study where there are only two classes, i.e. '*eruption*' and '*no eruption*', is further explorer in Chapter 8. Another common application of image classificaion in EO is to identify the type of cover of an image, as illustrated in Figure 6.8. In this case, where the goal is to identify if the image is '*urban area*', '*forest cover*' and '*water*', the training dataset will consist of **EO images**, and **tabular data**, where each image is assigned a vector containing information regarding its label.

6.5.1 Annotation tool for image classification – makesense.ai

Makesense.ai [10] is a free to use online tool that supports annotation of image (both for image classification and for object detection tasks). This tool does not require installation and it is accessible through the website: *makesense.ai*

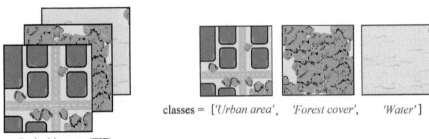

classes = ['*Urban area*',   '*Forest cover*',   '*Water*']

Optical images (TIF)

Figure 6.8   Example of training dataset for image classification: labels (in vector/tabular format) are assigned to each image

(see Figure 6.9). To annotate the EO dataset, the user can upload the dataset by dragging it into the middle selection or by selecting the image.

Once the images are uploaded, select the *Image Recognition* tab. The next step is to define the labels. Following the example in Figure 6.8, we choose '*Urban area*', '*Forest cover*' and '*Water*'. Pressing '*Start project*' will initiate the labelling task (see Figure 6.10).

The user can go through each image and select the correct label on the side panel (see Figure 6.11).

Once the labelling is over, these labels can be downloaded in multiple formats; in this case, they are downloaded in the CSV format (see Figure 6.12). The CSV format contains two columns with the name of the image and the corresponding label that the user has manually selected.

A further step might be necessary to complete the preparation of the training dataset in order to make it readable by the algorithm. This step could be,

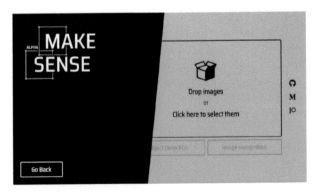

Figure 6.9   makesense.ai: online labelling software

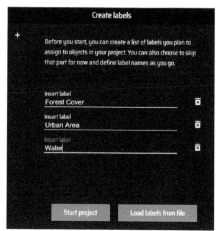

Figure 6.10   makesense.ai: defining the labels

for instance, to turn this exported CSV data into a vector, and finally creating the input data y. One possible approach is shown below, with a jupyter notebook and pandas. Depending on the algorithm and model, further structuring of this tabular data may be required (one vector encoding, for example).

6.6   Training dataset for semantic segmentation tasks

In a semantic segmentation task, also normally referred to as per-pixel classification, each pixel of an image is classified as belonging to a class. While image classification aims at telling if an image contains a forest in it for instance, semantic segmentation will allow to tell which pixel of the image belongs to the class *'forest'*. The training dataset for a semantic segmentation

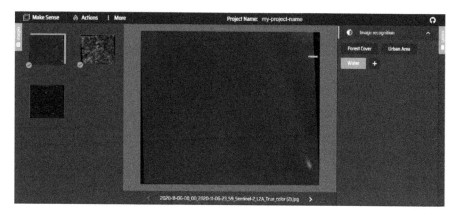

Figure 6.11   makesense.ai: defining the labels

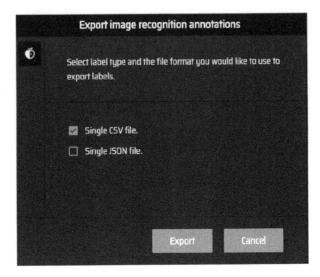

Figure 6.12   makesense.ai: exporting the labels

problem can be composed of a set of **images** and their **annotated masks** or a set of pixels collected from the images. Figures 6.13 and 6.14 show two different ways of creating a dataset for semantic segmentation tasks. In the example illustrated in Figure 6.13, the dataset will consist of the pixels or polygons of EO data and their corresponding labels. In the example shown in Figure 6.14, the training dataset consists of an entire image – here called as 'mask' consisting of the classes of each pixel (e.g. in Figure 6.13). A class value is assigned to each pixel, which can be represented with an integer or and RGB triplet.

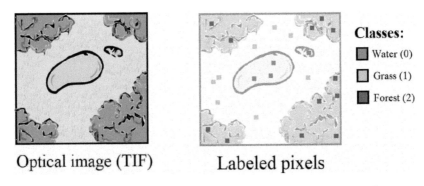

Figure 6.13   An example of a training dataset for an application aimed at classifying per pixel the land cover, collecting pixels from the image

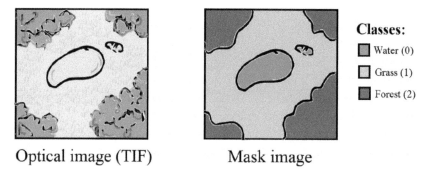

Optical image (TIF)   Mask image

*Figure 6.14   An example of a training dataset for an application aimed at
classifying per pixel, the land cover creating a mask from the image*

6.6.1   Annotation tool for semantic segmentation – Google Earth Engine online editor

GEE can also be used to annotate or label pixels (or polygons) and export them as training data (in case the user wants to develop the AI pipeline outside the GEE). In this example, the training data is collected by collecting pixels on one-month median values of Sentinel-2 data (just for the sake of simplicity.)

Sentinel-2 data composite and visualization

```
// Load Sentinel-2 TOA reflectance data.
var collection = ee.ImageCollection('COPERNICUS/S2')
    .filterDate('2020-08-01', '2020-08-31')
    // Pre-filter to get less cloudy granules.
    .filter(ee.Filter.lt('CLOUDY_PIXEL_PERCENTAGE', 20))
var composite = collection.median()

// Display the results.
Map.addLayer(composite,
             {bands: ['B4', 'B3', 'B2'],
              min: 0,
              max: 3000},
              'RGB')
```

In the upper left corner of the code editor visualisation panel, the user can select *Geometry Imports* and then add a *new layer* according to the number of classes of interest (Figure 6.15).

Before the start of sampling the authors recommend to first edit each of these layers. Hovering the mouse over each layer shows a gear icon. Clicking on it the user can *Edit layer properties. This will allow to set the label of each class and assign it a number* (Figure 6.16) *and change it to FeatureCollection.*

Once the classes are edited, the user can start collecting the pixels (with pins, rectangles or hand-drawn polygons) (Figure 6.17).

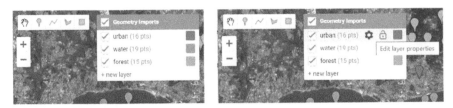

Figure 6.15   Creation of classes to be sampled

Once the placement of pins/polygons has been finalised, the user will need to sample the bands of the Sentinel-2 data that have the pin locations. The snippet code below shows how to merge the entire pins, and how to select the bands and property (in this case was named *class*). This step creates the actual training data.

Training data creation

```
// Merging all points in a single FeatureCollection.
var points =urban.merge(forest).merge(water);

// Select bands used for training
var bands = ['B2', 'B3', 'B4', 'B5', 'B6', 'B7'];

// Sample the composite to generate training data.
var training = composite.select(bands).sampleRegions({
  collection: points,
  properties: 'class',
  scale: 30
});
```

The last step consists of exporting the training data. On GEE a *FeatureCollection* can be exported as CSV, SHP (shapefile), GeoJSON, KML, KMZ or TFRecord.

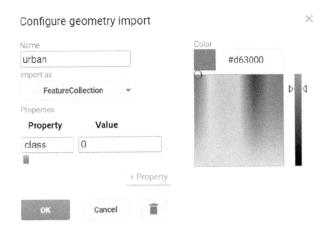

Figure 6.16   Editing class properties

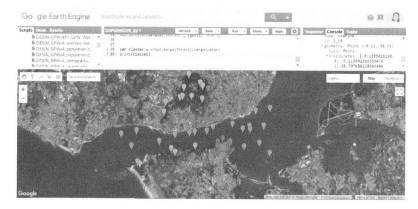

Figure 6.17   Picking pixels of different classes

Export Data

```
//Exporting the data
Export.table.toDrive({
  collection:training,
  description:'trainingDataName',
  folder:'GEE',
  fileFormat:'TFRecord' });
```

While the example below has shown how a training dataset for a semantic segmentation task by sampling pixels can be created, another widely used and perhaps more labour-intensive approach is the creation of a *mask* image as illustrated in Figure 6.14. The creation of a mask will require an annotation software, often called the 'brush' tool, that allows the user to manually 'paint' the whole picture with different classes. The authors recommend the users explore tools like LabelBox [11] a web-based platform that offers several ways to annotate such as points, lines, boxes, polygons and 'brush'. The output can be a JSON or CSV file (containing all annotations) or a PNG mask.

6.7   Training dataset for object detection tasks

The creation of a training dataset for an object detection (or object recognition) task could be more challenging (or more time-consuming) as it involves drawing a box – the ***bounding box***, from one corner to another of one or more objects in an image. As mentioned in Chapter 3, although mostly used for self-driving cars, object detection can also be used in Earth observation applications, where there are multiple targets in the image that we are interested in – like airplanes or ships, for example. The bounding boxes are thus used to classify the object of interest and to locate the object within the image. The

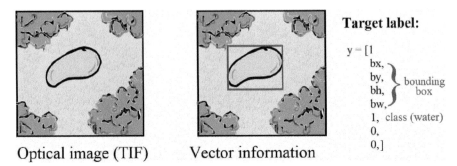

Optical image (TIF)   Vector information

Target label:

$y = [1$
$bx,$
$by,$ } bounding
$bh,$ } box
$bw,$
$1,$ class (water)
$0,$
$0,]$

Figure 6.18   Example of a training dataset for objection detection aimed at detecting water bodies

bounding box will be stored as vector information, providing information if the object exists; the coordinates are the centre of the rectangle, *bx, by*, as well as weight *bw* and height *bh* of the rectangle, and the class of the object. The annotated training set consists of pairs of images with their corresponding files containing information on the bounding boxes of that image as illustrated in Figure 6.18.

6.7.1   Annotation tool for semantic segmentation – LabelImg

LabelImg [12] is a free graphical image annotation tool that only supports bounding boxes. The annotation can be saved in the PASCALVOC XML format and it also supports the YOLO format. LabelImg can be downloaded at tzutalin.github. io/labelImg/ and then unzipped.

The authors describe how it can be installed using *Anaconda* (an example with *pip* has already been given). Anaconda is also widely popular and allows to create environments, preventing issues when installing libraries of different versions to other libraries already installed.

Python 3.8 is recommended, and as such, since it is needed to install some libraries, the authors would recommend creating a new environment and installing the new libraries there.

Installing libraries in a new environment

```
#Creates environment
conda create -n newEnvironment python=3.8
#Activates new environment
conda activate newEnvironment
conda install -c anaconda lxml
conda install -c anaconda pyqt
```

Once the necessary libraries to run LabelImg are installed, the user can open the software to start annotating images.

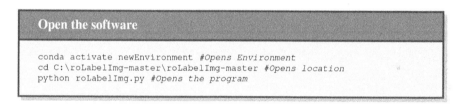

Figure 6.19   Image labelling software

Open the software

```
conda activate newEnvironment #Opens Environment
cd C:\roLabelImg-master\roLabelImg-master #Opens location
python roLabelImg.py #Opens the program
```

The images to be annotated are accessible on the left side of the interface of LabelImg (Figure 6.19), by clicking on the icons *Open* or *Open Dir*. In the following example the authors use a Sentinel-2 image, with the goal of generating a set of training airplanes. The bounding box can be created by clicking on *Create RectBox*, and should be drawn around the target – in this case the airplane (see Figure 6.20). Once the rectangle is created, it should be assigned to a class *'airplane'*. This procedure should be repeated for all the objects in the image. The output file will be saved separately for each image in the source folder.

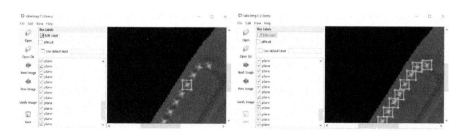

Figure 6.20   Creating bounding boxes with class 'plane' on Sentinel-2 data

6.8   Conclusions

As described earlier, there are several ways to access satellite data – and to create a training dataset for supervised learning. The way a training dataset is annotated will be determined by the type of ML task one needs to perform. Currently there is a growing number of free and open source annotation tools to support the creation of datasets; nevertheless it often requires an extra step to make the dataset ready to be used by an ML pipeline.

References

[1]   Schmitt M., Hughes L.H., Zhu X.X. 'The SEN1-2 dataset for deep learning in SAR-optical data fusion'. arXiv. 2019.

[2]   Liu Z., Yuan L., Weng L., Yang Y. 'A high resolution optical satellite image dataset for SHIP recognition and some new baselines'. *In International Conference on Pattern Recognition Applications and Methods*. 2017;**2**:324–31.

[3]   Cheng G., Han J., Lu X. 'Remote sensing image scene classification: benchmark and state of the art'. *Proceedings of the IEEE*. 2017;**105**(10):1865–83.

[4]   Van Etten A., Lindenbaum D., Bacastow T.M. 'Spacenet: a remote sensing dataset and challenge series'. arXiv. 2018.

[5]   European Space Agency (ESA). *Copernicus Open Access Hub [online]*. 2020. Available from https://scihub.copernicus.eu/.

[6]   European Space Agency (ESA). *Sentinel application platform (SNAP) [online]*. 2020. Available from http://step.esa.int/main/toolboxes/snap/.

[7]   European Space Agency (ESA). *SNAP tutorials [online]*. 2020. Available from http://step.esa.int/main/doc/tutorials/.

[8]   Google. *Google Earth Engine: JavaScript and Python guides [online]*. 2018. Available from https://developers.google.com/earth-engine/guides.

[9]   Sebastianelli A., Rosso M.P.D., Ullo S.L. *Auto-matic dataset builder for machine learning applications to satellite Imagery [online]*. 2019. Available from https://arxiv.org/abs/2008.01578.

[10]   Skalski P. *Make sense [online]*. 2019. Available from https://github.com/SkalskiP/make-sense/.

[11]   Wada K. *labelme: image polygonal annotation with Python [online]*. 2016. Available from https://labelbox.com/.

[12]   Tzutalin. *LabelImg. Git code [online]*. 2015. Available from https://github.com/tzutalin/labelImg.

Chapter 7

How to develop your network with Python and Keras

Silvia Liberata Ullo[1], Maria Pia Del Rosso[1], Alessandro Sebastianelli[1], Erika Puglisi[2], Mario Luca Bernardi[1], and Marta Cimitile[3]

This chapter introduces the reader to the implementation and training of simple deep learning model. The chapter includes examples that have been developed in Python, using the Keras library build on top of TensorFlow.

7.1 Introduction

7.1.1 TensorFlow and Keras

TensorFlow is a free open-source software library, developed by the Google Brain team, that provides a collection of workflows to develop and train models [1]. The name "TensorFlow" expresses the programming paradigm on which this library is implemented: the operations are performed on *tensors* (see Section 7.1.2) and the computations are expressed as *stateful dataflow graphs*, which means they are designed as directed graphs of data flowing between operations which are also able to remember preceding events and user interactions. The main features of TensorFlow are:

- a fast debug with Python tools
- the possibility to create dynamic models with Python control flow and multiple levels of abstraction
- support for custom and higher-order gradients
- flexibility and control with features like the Keras functional application programming interface (API)
- a good documentation, so it is easy to understand

[1]Engineering Department, University of Sannio, Italy
[2]Φ-Lab, European Space Agency (ESA), Italy
[3]Unitelma Sapienza University of Rome, Italy

Keras is an open-source high-level neural networks, library written in Python, capable of running on top of TensorFlow and other software [2]. It was developed as part of the research effort of project ONEIROS (Open-ended Neuro-Electronic Intelligent Robot Operating System) and its primary author and maintainer is François Chollet, a Google engineer. This library provides essential abstractions and building blocks for developing machine learning solutions with high iteration velocity. The main features of Keras are:

- a focus on user experience with a simple architecture that is readable and concise
- multi-backend and multi-platform
- support for convolutional and recurrent networks
- highly modular neural networks library written in Python
- allows for fast prototyping

In 2017, Google's TensorFlow team decided to support Keras in TensorFlow's core library, this is why TensorFlow's high-level APIs are based on the Keras API standard for defining and training neural networks. However, the two libraries have some important differences:

- Keras is an high-level API, while TensorFlow is a framework that offers both high- and low-level APIs.
- Keras is perfect for quick implementations, while TensorFlow is ideal for deep learning research and complex networks.
- Keras uses another API debug tool, while TensorFlow can use TensorBoard visualization tools for debugging.
- Keras is usually used for small datasets and low-performance models, whereas TensorFlow is used for large datasets and high-performance models.

In Section 7.3, some examples to learn how to develop a simple model using TensorFlow and Keras are provided. To run the examples, the reader should follow the installation guide in section LABEL:tensorflow_environment before; for a simpler use the authors suggest to work with Google Colaboratory (Section 7.2), which has a preinstalled version of TensorFlow.

7.1.2   Tensors

Tensors are immutable multidimensional arrays with a uniform type, called *dtype*, so it is not possible to update the contents of a tensor, but only to create a new one. In code block 7.1 some examples of tensors are shown.

Declaration and assignment of tensors   **7.1.2**

```
import tensorflow as tf

rank_0_tensor = tf.constant(9)
print(rank_0_tensor)

rank_1_tensor = tf.constant([0.4, 3.2, 6.1])
print(rank_1_tensor)

rank_2_tensor = tf.constant([[1, 2],
                             [3, 4],
                             [5, 6]], dtype=tf.float16)
print(rank_2_tensor)
```

Output   **7.1.2**

```
tf.Tensor(4, shape=(), dtype=int32)
tf.Tensor([2. 3. 4.], shape=(3,), dtype=float32)
tf.Tensor([[1. 2.]
           [3. 4.]
           [5. 6.]], shape=(3, 2), dtype=float16)
```

As it is possible to see, the first tensor is an int32 by default and represents a scalar; the second tensor is recognized as a float32 because the numbers assigned to it are float numbers and it represents a vector of size 3 (because it has three variables assigned); the last tensor is a float16, as specified at creation time, and represents a matrix composed of three rows and two columns. The *rank* of a tensor is the number of its dimensions, e.g. a scalar has rank 0, a vector has rank 1, and a matrix has rank 2.

Tensors can also have multiple dimensions or axes (see code block 7.1).

Declaration and assignment of a 3-axes tensor   **7.1.2**

```
rank_3_tensor = tf.constant([ [[0, 1, 2, 3, 4],
                               [5, 6, 7, 8, 9]],

                              [[10, 11, 12, 13, 14],
                               [15, 16, 17, 18, 19]],

                              [[20, 21, 22, 23, 24],
                               [25, 26, 27, 28, 29]],
                            ])
print(rank_3_tensor)
```

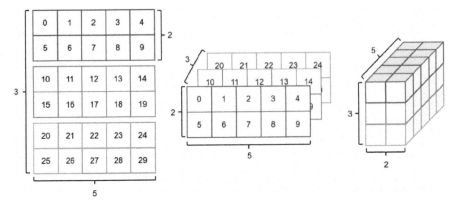

Figure 7.1   *Graphic representation of a three-axes tensor of shape (3,2,5)*

Output	7.1.2

```
tf.Tensor([ [[ 0  1  2  3  4]
             [ 5  6  7  8  9]]

            [[10 11 12 13 14]
             [15 16 17 18 19]]

            [[20 21 22 23 24]
             [25 26 27 28 29]]], shape=(3, 2, 5), dtype=int32)
```

The tensor shown in the code block 7.1 is a three-axes tensor of shape (3,2,5) that can be visualized as a group of three matrices composed of two rows and five columns, as it is possible to see in Figure 7.1.

It is possible to reshape a tensor using the tf.reshape operation. For example, a tensor of rank 2, which represents a matrix, can be *flattened* to a tensor of rank 1, so the elements of the matrix become the elements of a vector maintaining the row by row order.

More information about tensors or TensorFlow basics can be found in the TensorFlow Guide [3].

7.2   Google Colaboratory

Google Colaboratory, or Google Colab, is a free Jupyter Notebook environment that runs in the cloud and stores its notebooks on Google Drive [4]. The authors propose the use of Google Colab for three main reasons:

1. This platform allows you to start working on the application in a simple way, avoiding downloading and installing software. Google Colab does

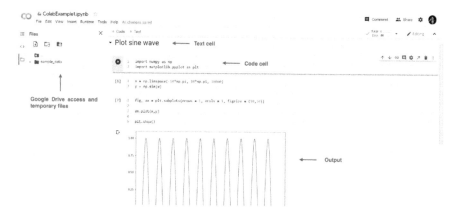

Figure 7.2   Google Colab GUI (Source [5])

not need to enable virtual environments and has almost all libraries related to data science and artificial intelligence (AI). This platform also allows you to install the missing libraries, and this aspect will be deepened in this chapter.

2. Google Colab makes graphics processing unit (GPU) available to the user for free, this aspect is of fundamental importance if you find yourself working with computationally expensive models.

3. The platform in question can interface in a very simple way with Google Drive. This feature allows the user to use datasets that occupy a lot of memory, avoiding uploading or downloading at every instance of the program.

Google Colab has also a paid plan called Colab Pro. This plan allows the user to get faster GPU, longer runtime, and extra RAM and memory.

As you can see from Figure 7.2, Google Colab has a graphical user interface (GUI) quite similar to Jupyter Notebook or Jupyter Lab. There are two main panels, one for organizing files and operating on them and one for coding and writing text.

As well as for Jupyter environments, the user can write text or notes in the Markdown language[a] and python code. Both the text and the code are organized in cells, each cell can be run separately, but global variables can be used as a normal. Each cell can produce an output. If the cell is text, the user should see only the formatted text, whereas if the cell is code, the user will see the code and its output as displayed below.

In order to run each cell, the user can click on the play button placed on the left of the cell or can click on the cell and press *Shift+Enter*.

---

[a]Markdown is a lightweight markup language with plain-text-formatting syntax, created in 2004 by JohnGruber with Aaron Swartz. Markdown is often used for formatting README files, for writing messages in online discussion forums, and for creating rich text using a plain text editor.

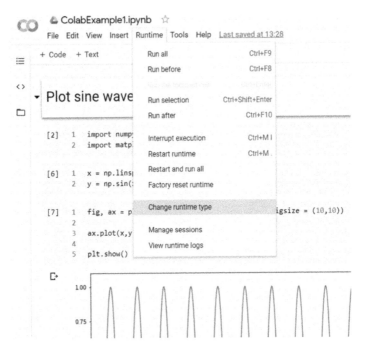

Figure 7.3   Connect the GPU (Step 1) (Source [5])

7.2.1   How to work with GPU?

As said before, one of the most useful and important features of Google Colab is the availability of free GPUs. In order to activate the GPU, or more specifically, in order to change the runtime type, the user should select the option *runtime* from the menu on top (see Figure 7.3).

After clicking on runtime, the user should select the option *change runtime type*. Then, as you can see from Figure 7.4, the user can select one of the three options:

- None, for using only the CPU
- GPU, for using the GPU
- TPU, for using the TPU (tensor processing unit[b])

Google Colab is offering two different GPU models since 2019: K80 and T4. However, you cannot select which of them you want due to the availability issues. The specifications for the GPU are described in Table 7.1.

---

[b]A TPU is an AI accelerator application-specific integrated circuit developed by Google specifically for neural network machine learning, particularly using Google's own TensorFlow software.

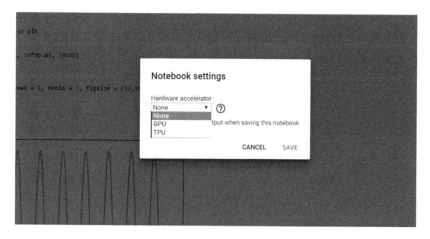

Figure 7.4   Connect the GPU (Step 2) (Source [5])

7.2.2   How to connect Google Drive?

Another important add-on that Colab offers is the Google Drive connectivity. Without connecting to Google Drive the user can upload files, but when the session expires the files are automatically removed. In order to avoid that kind of behavior, the user can link its own Google account, more precisely a Drive account. In this way the data can be loaded once on Drive and as many times as the user wants on Colab.

The linking procedure can be done by clicking on the folder panel (on the left) and then, as you can see from Figure 7.5, by selecting the Google Drive icon, a prompt appears. If the user selects the *CONNECT TO GOOGLE DRIVE* option, the Drive account is linked and all the files can be found in the *drive* folder. In the other case, as you can see from Figure 7.6, the user can use two lines of code to connect to Google Drive.

Another way to work with virtual environments consists in the use of Virtual Environment manager like conda or miniconda.

Table 7.1   Google Colab GPU specifications

Parameter	Value
GPU	Nvidia K80/T4
GPU memory	12 GB/16 GB
GPU memory clock	0.82 GHz/1.59 GHz
Performance	4.1 TFLOPS / 8.1 TFLOPS
Support mixed precision	No/yes
GPU release year	2014/2018
No. of CPU cores	2
Available RAM	12 GB (upgradable to 26.75 GB)
Disk space	358 GB

Figure 7.5   Connect Google Drive (Step 1) (Source [4])

7.3   Examples

In this section the authors propose some examples and scripts to guide the reader toward the implementation of a deep learning model. As discussed in the previous chapters the authors will focus essentially on deep learning applications. In order to gently introduce the user to tools and applications, in this section three simple applications will be presented. The applications are strictly linked to the ones that will be presented in the next chapters, which are also more focused on remote sensing.

The first step, for each example, consists of importing libraries for plotting data, for matrices handling and for developing machine learning model, and they are:

- NumPy: A library for the Python programming language, adding support for large, multidimensional arrays and matrices, along with a large collection of high-level mathematical functions to operate on these arrays.
- Matplotlib: A plotting library for the Python programming language and its numerical mathematics extension NumPy. It provides an object-oriented API for embedding plots into applications using general-purpose GUI toolkits like Tkinter, wxPython, Qt, or GTK+. There is also a procedural "pylab"

Figure 7.6   Connect Google Drive (Step 2) (Source [4])

How to set up a TensorFlow environment

In this section the authors explain how to create a safe environment in Windows and in Linux/MacOS in which TensorFlow code with or without the use of a GPU runs. Remember to check on the official TensorFlow website the system requirements so as to install the supported version of the programs.

Step 1 – Create a virtual environment

A virtual environment is a copy of an existing version of Python with the option to inherit existing packages. Python environments are important because they give the possibility to experiment different combinations of packages without affecting the main installation, moreover they are useful when a user has to work on a shared system and does not have permission to install packages in the main system. To create a virtual environment in Windows:

- Download and execute the latest Python installation package from: www.python.org/downloads/windows/. The version to install (Python 2 or Python 3) depends on the user end goal, since the authors suggest to install Python 3, this guide will use terminal commands for that version. Follow the instructions to install the package, choose the recommended installation options for a simple installation experience. Be sure during the installation to check the box for "add Python to PATH": this will adjust the System Environment Variables automatically so that Python can be launched from any command prompt, otherwise the user needs to add the Python PATH to System Environment Variables manually.
- Open the Windows Command Prompt and verify the successful Python installation typing *python3 -V*, the user should receive a message with the current installed version of Python.
- Usually pip package is already installed if the user is using Python 2 $\geq$ 2.7.9 or Python 3 $\geq$ 3.4, check its version typing the command *pip3 -V*. TensorFlow requires a pip version $\geq$ 19.0, if it is necessary an upgrade type the command: *python3 -m pip install –upgrade pip*. Otherwise, if the pip package is not installed, download get-pip.py[a] to a folder on the computer, open a command prompt and navigate to the folder containing the get-pip.py installer, then run the command *python3 get-pip.py*.
- Install with pip the virtualenv package [5] typing in the Windows Command Prompt *pip3 install virtualenv*.

[a]https://bootstrap.pypa.io/get-pip.py

(Continues)

- To create a virtual environment, the user must specify a path. For example to create a new one in the local directory called 'myvirtualenv', type the command *virtualenv myvirtualenv*.
- To activate the python environment run the command: *myvirtualenv/Scripts/activate*. The name of the virtual environment should appear in brackets in the terminal line. Now any python commands the user types will work with the virtual environment. To deactivate the virtual environment simply type *deactivate* in the prompt.

To create a virtual environment in Linux/MacOS:

- Open the Terminal program. This is usually found under Utilities or Accessories.
- Install Python typing the command *sudo apt-get install pythonX.X*, where X.X indicates the version to install (e.g. 3.7). Then, check the version typing: *python3 –version*.
- Install the pip package using the command *sudo apt install python3-pip* in the prompt and check the version with *pip3 –version*. TensorFlow requires a pip version ≥19.0, if it is necessary an upgrade type the command: *pip3 install –upgrade pip*.
- Install the virtualenv package [5] typing *pip3 install virtualenv*.
- To create a virtual environment, the user must specify a path. For example to create a new one in the local directory called 'myvirtualenv', type the command *virtualenv myvirtualenv*.
- To activate the python environment run the command: *source myvirtualenv/bin/activate*. The name of the virtual environment will appear in brackets in the terminal line. Now any python commands the user types will work with the virtual environment. To deactivate the virtual environment simply type *deactivate* in the prompt.

Step 2a – TensorFlow CPU-only

If the computer does not have a GPU, it is possible to install a CPU-only version of TensorFlow. To install it on Windows:

- Create or activate the virtual environment that will be the working area.
- Type the command: *pip3 install –user –upgrade tensorflow*.

(Continues)

- Once the installation is complete, verify it with the following command:

python
import tensorflow as tf
print (tf. __version__).
To install it on Linux/MacOS:

- Create or activate the virtual environment that will be the working area.
- Type the command: *pip3 install –upgrade tensorflow.*
- Once the installation is complete, verify it with the following command:

python
import tensorflow as tf
print (tf. __version__).

Step 2b.1 – TensorFlow GPU for NVIDIA GPUs

To enable the use of the GPU on computers with an NVIDIA GPU, it is necessary to install CUDA [6] and the cuDNN [7] version compatible with the CUDA version chosen. The procedure using Windows is:

- Download and install Visual studio from the website: https://visualstudio.microsoft.com/it/downloads/ choosing the version that best matches with the end goal. This is a prerequisite for the installation of Nvidia CUDA Toolkit. After the installation it is necessary to reboot the computer.
- Check the list on this website: https://developer.nvidia.com/cuda-gpus to see if the GPU on the computer is a CUDA-enabled GPU.
- Visit this website: www.tensorflow.org/install/source#gpu to take note of which combination of CUDA, cuDNN, and TensorFlow versions are better to install.
- Go to the website: https://developer.nvidia.com/cuda-toolkit-archive click on a compatible CUDA toolkit and follow the instructions to install it by selecting the parameters which describe the target platform used.
- Check if the installation was successful by typing in the Command Prompt window the command *nvcc -V*, the output should show the CUDA version installed.
- Go to the website: https://developer.nvidia.com/cudnn-download-survey it is necessary to login and create an account to access

(Continues)

to the list of available cuDNN software. Download the cuDNN version that best matches with the CUDA version installed. Unzip the downloaded archive, navigate into the unzipped directory, and copy the following files into the CUDA Toolkit directory: (1) Copy cuDNN_directory/cuda/bin/cudnn*.dll to "C:/ Program Files/NVIDIA GPU Computing Toolkit/CUDA/vxx.x/ bin". (2) Copy cuDNN_directory/cuda/include/cudnn*.h to "C:/ Program Files/NVIDIA GPU Computing Toolkit/CUDA/vx.x/ include". (3) Copy cuDNN_directory/cuda/lib/x64/cudnn*.lib to "C:/Program Files/NVIDIA GPU Computing Toolkit/CUDA/ vx.x/lib/x64".

- It is necessary to update the environment variables. Open Control Panel – System and Security – System – Advanced System Settings and click *Environment Variables* at the bottom of the window. In the new window and in the *System variables* pane, select the *Path* variable and click *Edit*. There should be two CUDA entries already listed, but if they are not present then add them manually: "C: Program_Files NVIDIA_GPU_Computing_Toolkit CUDA vxx.x bin" "C: Program_Files NVIDIA_GPU_Computing_Toolkit CUDA vxx.x libnvvp".
- Now it is possible to install TensorFlow with GPU support. Open the Command Prompt as an Administrator, start the virtual environment, and type *pip3 install –upgrade tensorflow-gpu*.
- Test if everything is correctly installed by typing in the terminal:

python
import tensorflow as tf
tf.test.is_built_with_cuda()
tf.test.is_gpu_available(cuda_only=False,
min_cuda_compute_capability=None).

- If the outputs are "True" the system is ready to be used.

The procedure using Linux/MacOS is:

- Check the list on this website: https://developer.nvidia.com/cuda-gpus to see if the GPU on the computer is a CUDA-enabled GPU.
- Type the command *nvidia-settings* in a terminal to check what is the graphics card driver version in use.
- Visit this website: https://docs.nvidia.com/deploy/cuda-compatibility/index.html#binary-compatibility__table-toolkit-driver to identify which version of CUDA is compatible with the card.

(Continues)

- Visit this website: www.tensorflow.org/install/source#gpu to take note of which cuDNN and TensorFlow versions are better to install based on the most compatible CUDA version.
- Go to the website: https://developer.nvidia.com/cuda-toolkit-archive click on a compatible CUDA toolkit and follow the instructions to install it by selecting the parameters which describe the target platform used.
- Check if the installation was successful by typing in a terminal the command *nvcc -V*, the output should show the CUDA version installed.
- Go to the website: https://developer.nvidia.com/cudnn-download-survey it is necessary to login and create an account to access to the list of available cuDNN software. Download the cuDNN version that best matches with the CUDA version installed. The procedure to install from a Tar file is now described: extract the file launching the following command in the terminal *tar -xvzf cudnn-X.X-linux-x64-vX.X.X.XX.tgz* The value of the X depends on the version downloaded. Next, copy the extracted files to the CUDA installation folder using these commands:
- *sudo cp cuda/include/cudnn.h /usr/lib/cuda/include/ sudo cp cuda/lib64/libcudnn* /usr/lib/cuda/lib64/*
- and set the file permissions of cuDNN:
- *sudo chmod a+r /usr/lib/cuda/include/cudnn.h /usr/lib/cuda/lib64/libcudnn*.*
- It is necessary to add the CUDA environment variables in the system path. Open the .bashrc file using a text editor and append the following lines:
- *export   LD_LIBRARY_PATH=/usr/lib/cuda/include:$LD_LIBRARY_PATH*
- *export   LD_LIBRARY_PATH=/usr/lib/cuda/lib64:$LD_LIBRARY_PATH.*
- Then load the exported environment variables by running *source /.bashrc*.
- Now it is possible to install TensorFlow with GPU support. Open the terminal, start the virtual environment, and type *pip3 install tensorflow* (with the introduction of TensorFlow 2.0 it is no more necessary to specify the version with/without GPU that the user wants to install) or *pip3 install tensorflow-gpu*.
- Test if everything is correctly installed by typing in the terminal:

  python
  import tensorflow as tf

(Continues)

tf.test.is_gpu_available(cuda_only=False, min_cuda_compute_capability=None).

- If the output is "True" the system is ready to be used.

Step 2b.2 - Tensorflow GPU for AMD GPUs

If the computer has an AMD GPU then it is not possible to install the CUDA Toolkit, used by TensorFlow GPU, because it is proprietary to NVIDIA. In case the user has the possibility to write the code using only Keras library, then it is sufficient to install the PlaidML framework[b] [8] and add the following two lines to the code:

import os
os.environ["KERAS_BACKEND"] = "plaidml.keras.backend".
Otherwise, if the user has the requirement to use the TensorFlow library:

- In Windows: up to now there is no solution, the authors suggest to use Google Colab.
- In Linux: it is possible to install the AMD ROCm Platform (instructions available on website: https://rocmdocs.amd.com/en/latest/Installation_Guide/Installation-Guide.html) [9] and, then, the TensorFlow-rocm version (instructions available on website: https://rocmdocs.amd.com/en/latest/Deep_learning/Deep-learning.html).

[b]Instructions available on website: https://plaidml.github.io/plaidml/docs/install.html

---

interface based on a state machine (like OpenGL), designed to closely resemble that of MATLAB®, though its use is discouraged. SciPy makes use of Matplotlib.

- Keras: An API designed for humans, not machines. Keras follows best practices for reducing cognitive load: it offers consistent and simple APIs, minimizes the number of user actions required for common use cases, and provides clear and actionable error messages. It also has extensive documentation and developer guides.

7.3.1   Noise filtering

The first application proposed involves the theme of removing noise from a signal. More precisely, the idea is to develop a model able to filter noise, with normal distribution, from sinusoidal signals.

7.3.1.1 Step 1 – Define a generator for the dataset

For this simple example the dataset is created online. A common practice in deep learning is the use of generators. Basically a generator is composed of an infinite while loop and of the yield keyword. The generator is used to avoid memory problems, in fact the dataset is not loaded all at once, but batch by batch. In this case the concept is similar, for each iteration of the infinite while loop, a new batch of input and output data is produced.

As you can see in code block 7.2, the variable x_train and y_train represent respectively the input and the desired output. At the beginning both are initialized to zero and then during the first for loop, both are set to the values of sine of . Then, during the second for loop, normal noise is added to each element of the x_train vector.

Define a generator for the dataset   **7.3.1.1**

```python
def dataset_generator(batch_size, time_size):
    x_train = np.zeros((batch_size, time_size))
    y_train = np.zeros((batch_size, time_size))
    t = np.linspace(-3*np.pi, 3*np.pi, time_size)

    while True:
        for i in range(batch_size):
            # Create random sine waves with
            m = np.random.randint(1, 4)
            d = np.random.randint(0, 360)

            x_train[i,...] = np.sin(m*t + d)/2
            y_train[i,...] = np.sin(m*t + d)/2
            # Add noise to x_train
            for dt in range(time_size):
                noise = np.random.normal(0, 1, size = 1)
                x_train[i, dt] = x_train[i, dt] + 0.3*noise/2

        yield x_train, y_train
```

The generator produces, in this particular case, both the input and the output. In Figure 7.7 you can see an example of input and output produced by the generator.

7.3.1.2 Step 2 – Build a model

The natural step after the implementation of dataset is the design of a model. During this phase the user can build a network completely new or based on common architecture that can be found in the state of the art [10–19]. In this case the model is composed of four fully connected layers, with a dimension that depends on the variable time_size (that represents the length of each sine wave) and the tanh as activation function (see code block 7.3).

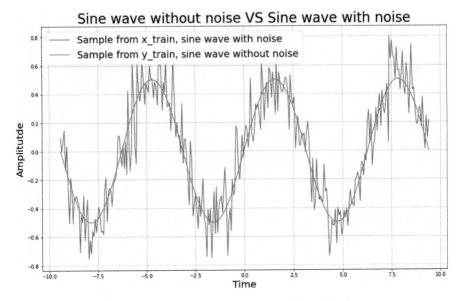

Figure 7.7   Example of input and output produced by the generator

Build a model	7.3.1.2

```
time_size =  x_train.shape[1]

model = Sequential()
model.add(Dense(time_size,
         input_dim=time_size, activation='tanh'))
model.add(Dense(time_size//2, activation='tanh'))
model.add(Dense(time_size//4, activation='tanh'))
model.add(Dense(time_size, activation='tanh'))
```

Once the structure is ready, the model must be compiled. For simplicity, this process consists in assigning a loss function and an optimizer. For this example, as you can see in code block 7.3, the loss function is the mean absolute error (MAE) and the optimizer is Adam.

Compile the model	7.3.1.2

```
model.compile(loss='mae', optimizer='adam')
```

As for the model structure, the choice of the loss function and of the optimizer can be driven by experience, by tuning or by studies from the state of the art. Obviously there exist rules and common practices that help the choice of these parameters.

7.3.1.3   Step 3 – Train and test the model

The function for the training of the model, described in this section, is *fit genera-tor*. This function needs a data generator, the number of epochs, and the step for each epoch (typically length of the dataset divided by the batch size).

Train the model	7.3.1.3

```
history = model.fit_generator(
    dataset_generator(16, time_size),
    steps_per_epoch=4000//16,
    validation_data=dataset_generator(16, time_size),
    validation_steps=4000//16,
    epochs=20
)
```

In order to test the model, as you can see in code block 7.4, new data are needed. The generator produces randomly/new data when the *next* function is called, but in order to apply this function the generator must be converted into an iterator using the *iter* function. The new data can finally be used to predict the output through the model.

Test the model	7.3.1.3

```
# Generate new data
generator = dataset_generator(10, time_size)
iterator = iter(generator)
x_train, y_train = next(iterator)
# Predict the output using the model
 prediction = model.predict(x_train)
```

The *prediction* vector contains the filtered sine waves. In Figure 7.8 the reader can see the results, compared also with the ground truth (y_train) and with the noisy inputs (x_train).

As you can see in Figure 7.8, the task has been solved using a very simple architecture. Obviously the results can be improved by tuning the parameters, by modifying the network, etc. The authors leave this exercise to the interested reader and invite her/him to post questions and its solution on the Git-Hub page [20], where they can also find the Jupyter Notebook.

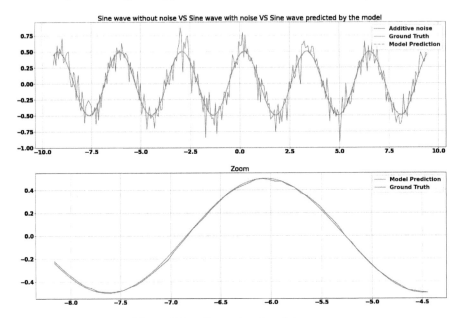

Figure 7.8   Example of model output

7.3.2   Classification

The second application proposed involves the theme of objects classification. The goal of this exercise is to implement a model that can classify six different types of objects in an image corrupted by noise.

7.3.2.1   Step 1 – Define a generator for the dataset

Also in this case the dataset is created online. Here, the object, a colored square, is placed randomly in an image of 32 × 32 pixels size. The class defines the color of the object. During the for loop, in code block 7.5, each sample of the batch is filled with an image in which there is only one object of a random class, placed in a random position. The size of the object is fixed in this case. At the end of the if–else block, which assigns the class, the noise is added to each sample of the batch. The generator returns both the images and a vector that indicates the class for each image.

```python
def dataset_generator(batch_size = 16):
  img_shape = (32, 32, 3)
  object_size = 8
  while True:
    x_train = np.zeros((batch_size,
                        img_shape[0], img_shape[1], img_shape[2]))
    # The classes are represented by the colors
    y_train = np.zeros((batch_size, 6))

    for b in range(batch_size):
      selector = np.random.randint(0, high = 6, size = 1)
      x = np.random.randint(object_size,
                            high = img_shape[0] - object_size)
      y = np.random.randint(object_size,
                            high = img_shape[1] - object_size)

      # Create and object randomly (random position and random class)
      if selector == 1:
        # Red object
        x_train[b, x:x+object_size, y:y+object_size, 0] = 0.6
        y_train[b, 1] = 1.0
      elif selector == 2:
        # Green object
        x_train[b, x:x+object_size, y:y+object_size, 1] = 0.6
        y_train[b, 2] = 1.0
      elif selector == 3:
        # Blue object
        x_train[b, x:x+object_size, y:y+object_size, 2] = 0.6
        y_train[b, 3] = 1.0
      elif selector == 4:
        # Yellow object
        x_train[b, x:x+object_size, y:y+object_size, 0] = 0.5
        x_train[b, x:x+object_size, y:y+object_size, 1] = 0.5
        y_train[b, 4] = 1.0
      elif selector == 5:
        # Fuchsia object
        x_train[b, x:x+object_size, y:y+object_size, 0] = 0.5
        x_train[b, x:x+object_size, y:y+object_size, 2] = 0.5
        y_train[b, 5] = 1.0
      else:
        # No object
        x_train[b, ...] = 0
        y_train[b, 0] = 1.0

      # Add noise to the batches
      x_train[b, ...] = x_train[b, ...] + noise()
      x_train[b, ...] = np.clip(x_train[b, ...], 0.0, 1.0)

    yield x_train, y_train
```

As you can see in code block 7.5, the noise function add both a background noise, with a uniform distribution, and a fake object with a random color. This function has the purpose of increasing the robustness of the model, which became able to classify objects in noisy environments. This is a common practice in machine learning [21].

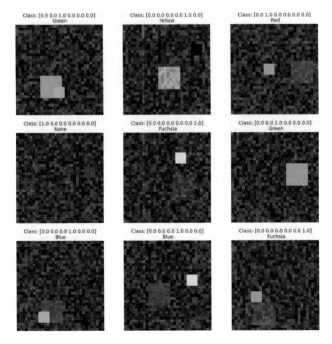

Figure 7.9   Example of input and output produced by the generator

```
def noise(noise_power=0.3):
    x = np.random.randint(4, high = 32 - 4)
    y = np.random.randint(4, high = 32 - 4)

    noise = np.random.random((32, 32, 3))*noise_power
    noise[x:x+4, y:y+4, ...] = np.random.randint(0, 2, size = 3)

    return noise
```

An example of an output produced by the generator is shown in Figure 7.9. As you can see, the title of each image represents the vector with the reference to the class (y_train), for simplicity the title also contains the name of the class. In the image (x_train), as said before, you can see the real object (the square with a "transparent" color), the fake object (the small square), and the background noise.

7.3.2.2   Step 2 – Build a model

As mentioned before the structure of the model can be driven by different factors, in this case the model is composed of different dense layers, with a decreasing size and the ReLU as activation function (except for the last layer). As you can see in code block 7.6, before the first dense layer there is a flatten layer. Since the input is a 3D matrix of size $32 \times 32 \times 3$ and the dense layer is expecting a 1D vector, the flatten layer is necessary. This layer unrolls the matrix in a 1D vector.

Build a model	7.3.2.2

```
model = Sequential()
model.add(Flatten(input_shape=(32, 32, 3)))
model.add(Dense(512, activation = 'relu'))
model.add(Dense(128, activation = 'relu'))
model.add(Dense(64, activation = 'relu'))
model.add(Dense(16, activation = 'relu'))
model.add(Dense(6, activation = 'softmax'))
```

In this case, as you can see in code block 7.6, the model is compiled using as loss function the binary cross-entropy, the Adam as optimizer. As extra metrics, commonly for classification tasks, the authors selected the accuracy.

Compile the model	7.3.2.2

```
model.compile(loss='binary_crossentropy', optimizer='adam',
              metrics=['accuracy'])
```

7.3.2.3   Step 3 – Train and test the model

The model will be trained with the function fit_generator, for 10 epochs. This function uses generators, for both training and validations sets. Since the generator is producing an infinity of samples, using steps_per_epoch and validation_step, the user can define an arbitrary size for the dataset.

Train the model	7.3.2.3

```
history = model.fit_generator(
    dataset_generator(16),
    steps_per_epoch=1000//16,
    validation_data=dataset_generator(8),
    validation_steps=100//8,
    epochs = 10
)
```

To test the model the user can run the code in code block 7.7.

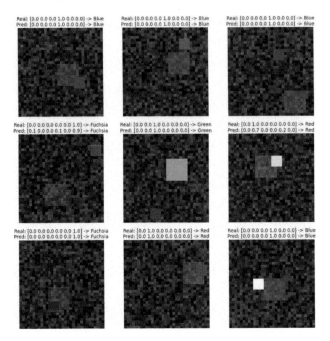

Figure 7.10   Example of model output

```
# Generate new data
generator = dataset_generator(9)
iterator = iter(generator)
x_train, y_train = next(iterator)
# Predict the output using the model
prediction = model.predict(x_train)
```

The *prediction* vector contains the class of the object for each image. In Figure 7.10 the reader can see some results, compared also with the ground truth (y_train).

For classification problems, it is very common to calculate the confusion matrix. In the field of machine learning and specifically the problem of statistical classification, a confusion matrix is a specific table that allows visualization of the performance of a machine learning algorithm. Each row of the matrix represents the instances in a predicted class while each column represents the instances in an actual class (or vice versa). The name derives from the fact that this matrix is used to highlight if the model is confusing two classes.

As you can see in Figures 7.10 and 7.11, the task has been solved using a very simple architecture. Obviously the results can be improved by tuning the parameters, by modifying the network, etc. The authors leave this exercise to the interested

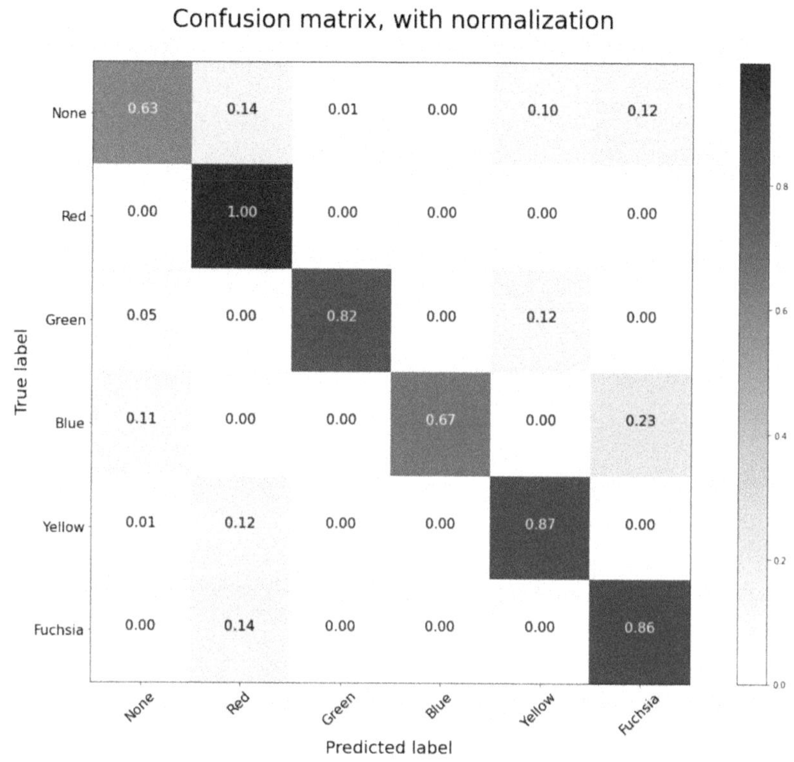

Confusion matrix, with normalization

Figure 7.11   Confusion matrix

reader and invite her/him to post questions and its solution on the Git-Hub page [14], where they can also find the Jupyter Notebook.

7.3.3   Detection

The theme of the third application is the object detection. The goal of this exercise is to implement a model able to detect simple objects in a noisy environment.

7.3.3.1   Step 1 – Define a generator for the dataset

Similar to the previous cases, the dataset is generated online through an ad hoc function. In this case the generator produces as output an image of 32 × 32 pixels, in which there is a rectangular object randomly placed, and a vector that contains the coordinates of the top-left and bottom-right corners. Also in this case there is a noisy background, which should help the model in being more robust. Alike the other cases (see generator in code block 7.8), the process of generation involves an endless loop and the keyword yield. The infinite loop contains a for loop used to generate a batch of samples at each iteration.

Define a generator for the dataset　　　　　　　　　　　　7.3.3.1

```
def dataset_generator(batch_size = 16):
  img_shape = (32, 32, 3)

  while True:
    x_train = np.zeros((batch_size, img_shape[0],
                  img_shape[1], img_shape[2]))
    y_train = np.zeros((batch_size, 4))

    for b in range(batch_size):
      object_width = int(np.random.randint(4, high = 12, size = 1))
      object_height = int(np.random.randint(4, high = 12, size = 1))
      #object_height = object_width

      x = np.random.randint(object_width + 1,
        high = img_shape[0] - object_width - 1)
      y = np.random.randint(object_height + 1,
        high = img_shape[0] - object_height - 1)

      x_train[b, x:x+object_width, y:y+object_height, 1] = 1
      x_train[b, ...] = x_train[b, ...]+np.random.random(img_shape)*0.2

      x_train = np.clip(x_train, 0.0, 1.0)

      y_train[b, 0] = x
      y_train[b, 1] = y
      y_train[b, 2] = x + object_width
      y_train[b, 3] = y + object_height

    yield x_train, y_train
```

An example of output produced by the generator is shown in Figure 7.12, in this case the title of each sub-figure shows the coordinates of the top-left and bottom-right corners. As you can see the size of each object is randomly selected.

7.3.3.2   Step 2 – Build a model
Also in this case the model, in code block 7.9, is a sequential model composed of multiple dense layers.The flatten layer is likewise used to unroll the image in a vector. The last layer of the network has a size of 4, which indicates the size of the expected output: the x and y coordinates for the top-left and bottom-right corner.

Build a model　　　　　　　　　　　　　　　　　　7.3.3.2

```
model = Sequential()
model.add(Flatten(input_shape=(32, 32, 3)))
model.add(Dense(512, activation = 'relu'))
model.add(Dense(512, activation = 'relu'))
model.add(Dense(128, activation = 'relu'))
model.add(Dense(64,  activation = 'relu'))
model.add(Dense(32,  activation = 'relu'))
model.add(Dense(4))
```

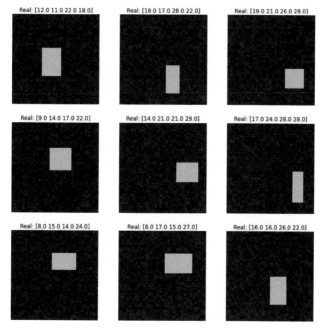

Figure 7.12   Example of input and output produced by the generator

As you can see in code block 7.9, in this case the model is compiled with the MAE as the loss function as Adam as the optimizer.

Build a model	7.3.3.2

```
model.compile(loss='mae', optimizer = 'adam')
```

7.3.3.3   Step 3 – Train and test the model

The training (code block 7.10) and the testing (code block 7.10) of the model can be done using the same method presented before.

Train the model	7.3.3.3

```
history = model.fit_generator(
    dataset_generator(16),
    steps_per_epoch=4000//16,
    validation_data=dataset_generator(8),
    validation_steps=400//8,
    epochs = 100
)
```

Test the model	7.3.3.3

```
# Generate new data
generator = dataset_generator(9)
iterator = iter(generator)
x_train, y_train = next(iterator)
# Predict the output using the model
 prediction = model.predict(x_train)
```

The *prediction* vector contains the predicted coordinates for the top-left and bottom-right corner of the object. These values are used to draw a bounding box

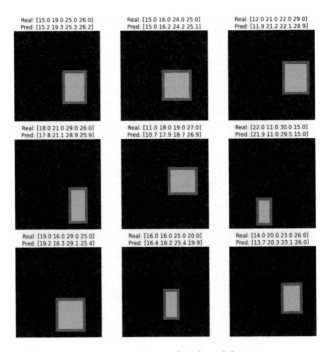

Figure 7.13   Example of model output

around the object. In Figure 7.13 you can see an example of a detected object. The title of each object shows the real positions compared with the predicted ones.

Also in this case the results can be improved by tuning the parameters, by modifying the network, etc. The authors leave this exercise to the interested reader and invite her/him to post questions and its solution on the Git-Hub page [14], where they can also find the Jupyter Notebook.

7.4 Conclusions

In conclusion, some example/tutorial for the implementation, training, and testing of neural networks have been presented in this chapter. Furthermore, in this chapter a development platform has been presented, alternatively the procedure for installing the development libraries on the personal computer has been described.

The following chapters will present the applications described in this chapter, but applied to satellite data.

References

[1] Team GB. *TensorFlow [online]*. Available from www.tensorflow.org/ [Accessed September 2020].

[2] Keras C.F. Keras. 2017. Available from https://keras.io/ [Accessed September 2020].

[3] Team GB. *TensorFlow Guide [online]*. Available from www.tensorflow.org/ guide [Accessed September 2020].

[4] Google. *Google Colaboratory [online]*. Available from https://colab.research. googl e.com/notebooks/intro.ipynb [Accessed September 2020].

[5] Virtualenv package [online]. Available from https://virtualenv.pypa.io/en/lat- est/index.html [Accessed September 2020].

[6] *CUDA toolkit, NVIDIA developer [online]*. 2012. Available from https://de- veloper.nvidia. com/cuda-toolkit [Accessed September 2020].

[7] *NVIDIA cuDNN, NVIDIA developer [online]*. Available from https://devel- oper.nv idia.com/cudnn [Accessed September 2020].

[8] *PlaidML framework [online]*. Available from https://github.com/plaidml/ plaidml [Accessed September 2020].

[9] *AMD ROCm Platform [online]*. Available from https://developer.nv idia.com/ cudnn [Accessed September 2020].

[10] Redmon J., Farhadi A. 'Yolo V2.0. Cvpr2017'. 2017.

[11] Redmon J., Farhadi A. 'YOLO v.3. Tech report'. 2018.

[12] Joseph R., Divvala S., Girshick R.,*et al*. 'YOLO) you only look once. Cvpr'. 2016.

[13] Yuan Z.-W., Zhang J. 'Feature extraction and image retrieval based on AlexNet'. *Eighth International Conference on Digital Image Processing (ICDIP 2016). International Society for Optics and Photonics*. 2016;**10033**(100330E).

[14] Hassan Mul. 'VGG16 – convolutional network for classification and detection'. *Neurohive. Dostopno na: https://neurohive. io/en/popular-networks/vgg16/[10. 4. 2019]*. 2018.

[15] Badrinarayanan V., Kendall A., Cipolla R. 'SegNet: a deep convolutional encoder-decoder architecture for image segmentation'. *IEEE transactions on pattern analysis and machine intelligence*. 2017;**39**(12):2481–95.

[16] Szegedy C., Liu W., Jia Y., *et al.* GoogLeNet. *Proceedings of the IEEE Computer Society Conference on Computer Vision and Pattern Recognition*; 2014.

[17] Ronneberger O., Fischer P., Brox T. U-net: Convolutional networks for biomedical image segmentation. International Conference on Medical image computing and computer-assisted intervention; Munich Germani, October 5-9; 2015.

[18] Zhou Z., Siddiquee R., Mahfuzur M.,*et al.* UNet++: redesigning skip connections to exploit multiscale features in image segmentation. *IEEE Transactions on Medical Imaging*; 2020.

[19] Zhou Z., Siddiquee R., Mahfuzur M.,*et al.* 'Unet++: a nested u-net architecture for medical image segmentation'. *Lecture Notes in Computer Science (including subseries Lecture Notes in Artificial Intelligence and Lecture Notes in Bioinformatics)*; 2018.

[20] Sebastianelli A. *Git-Hub page for code examples [online]*. Available from https://github.com/Sebbyraft/Code-Artificial-IntelligenceApplied-toSatellite-based-RemoteSensingData-for-EarthObservation [Accessed September 2020].

[21] He Z., Rakin Z., Fan Z. Parametric noise injection: Trainable randomness to improve deep neural network robustness against adversarial attack. *Proceedings of the IEEE/CVF Conference on Computer Vision and Pattern Recognition*; 2019. pp. 588–97.

Chapter 8

A classification problem

Alessandro Sebastianelli[1], Maria Pia Del Rosso[1], Chiara Zarro[1], Diego Di Martire[2], Mariano Di Napoli[2], Dario Spiller[3], Artur Nowakowski[4], Rogerio Bonifacio[5], Do-Hyung Kim[6], and Manuel Garcia-Herranz[6]

In this chapter, the problem of image classification, target detection, and objects segmentation is addressed. The chapter is full of case studies, in which both the dataset implementation procedure and the neural network architecture are presented.

8.1 Introduction

The aim of this chapter is to introduce the reader to a different approach for remote sensing data classification and detection, and also to introduce the processing chain typically adopted to solve real problems.

8.2 Case study A: volcanic eruptions binary classification

The aim of this case study is the implementation of a deep learning model for the binary classification of volcanic eruptions. Since no dataset was found for this specific tasks, a dataset was built using a catalog, containing the volcanic events, and the Satellite Data Downloader Tool presented in [1].

8.2.1 The catalog

The catalog used for the creation of the dataset is the Volcanoes of the World (VOTW) catalog by the Global Volcanism Program of the Smithsonian Institution, a

[1]Engineering Deparment, University of Sannio, Italy
[2]University of Naples Federico II, Italy
[3]Join ASI - Φ-Lab, Italian Space Agency (ASI) and European Space Agency (ESA), Italy
[4]Φ-Lab, European Space Agency (ESA), Italy
[5]World Food Program, Italy
[6]UNICEF, USA

Table 8.1   Sample from the catalog

Start time	Volcano name	Latitude	Longitude
June 26, 2019	Ulawun	−5.050	151.330
June 24, 2019	Ubinas	−16.355	151.330
June 22, 2019	Raikoke	48.292	153.250
June 11, 2019	Piton de la Fournaise	−21.244	55.708
June 1, 2019	Great Sitkin	52.076	−176.130

catalog of Holocene and Pleistocene volcanoes, and eruptions from the past 10,000 years. It is available at link: http://volcano.si.edu/database/search_eruption_results. cfm. An example of information in the dataset is reported in Table 8.1.

Since the goal is to create a set of images, the only useful information are the starting date, the coordinates, and the volcano name, so these information were extracted and stored apart. In this first phase only the volcanic eruptions since 1999 have been selected as it is the first useful date to download satellite multispectral images (Landsat 7 was launched that year). In the second phase only the ones that happened after the year of launch of the specific satellite will be considered, e.g. 2015 for Sentinel-2.

8.2.2   Data preparation and manipulation

The data, downloaded with the tool mentioned above, undergoes to a postprocessing procedure. The aim of these procedures is to combine infrared bands with RGB (Red-Green-Blue) bands. These bands are essential because usually the color of the volcanic lava, after an initial state, is dark and not visible from the space but its temperature is equally very high. In order to highlight this feature, a temperature-sensitive band, like the infrared ones, have been used.

The formula used for the band merging is given by (8.1):

$$\text{RED} = B4 \cdot 2.5 + \max(0, B12 - 0.1) \qquad (8.1a)$$

$$\text{GREEN} = B3 \cdot 2.5 + \max(0, B11 - 0.1) \qquad (8.1b)$$

$$\text{BLUE} = B2 \cdot 2.5 \qquad (8.1c)$$

In this case, the red and green bands are modified by adding respectively $B12$ and $B11$ information (two infrared bands both for Sentinel-2 and Landsat satellites): a red/yellow spot appears where there is a pixel with a higher temperature. The multiplicative factor is used only for scaling. In practice, the infrared bands change the red and green bands, so that the heat information is highlighted and visible to the human eye. In this way it was possible to create a quantitatively correct dataset, since during labeling the eruptions were easily distinguishable from non-eruptions.

In Figure 8.1, the difference between a simple RGB image and the processed one is shown.

Figure 8.1   True color image vs. infrared highlights image

In order to make the images more bright and clearly recognizable by the human vision, another transformation have been made on the dataset. In Table 8.2, the values chosen for the image correction are shown.

This settings resulted from processing the Mayon September 2, 2018, patch, equalizing the histogram for this patch. All others have inherited these settings.

8.2.3   Dataset expansion

Since the task is the binary classification, a set of images with no eruptions has been downloaded and added to the dataset. To obtain the best results, the no-eruption dataset should be as various as possible, containing a large amount of different scenarios. Following this idea these types of data have been downloaded in the same way as the eruption images:

- volcanoes images with no eruption happening
- cities images
- mountains images
- partly cloudy images with different shapes of clouds

Table 8.2   Image correction parameters

Setting name	Setting value
Exposure	+0.48
Contrast	+12
Highlights	−47
Shadows	+100
Whites	+55
Blacks	−29
Clarity	0
Vibrance	+14
Saturation	+5

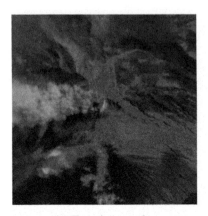

(a) Eruption smoke   (b) Volcano surrounded by clouds

Figure 8.2   Smoke (a) VS cloud (b)

- completely random images

It is really important to include cloudy images to distinguish from eruption clouds and smoke that have their proper shapes (see Figure 8.2).

The same type of preprocessing was also performed for these new data. In fact, for the deep learning context, it is necessary that the dataset is as homogeneous as possible. In fact, it is known that a neural network model is very sensitive to variations in the distribution of the input data.

The final dataset contains 260 images for the class "eruption." and more than 1500 for the class "no eruption." Due to the type of event analyzed, the dataset appears to be unbalanced. In fact, getting an image with a current eruption is a fairly rare event. This can be a problem for training the model and will be better addressed in the next section.

A set of images taken from the dataset is shown in Figure 8.3: the first and the second images show two volcanic eruption with lava rivers; the third and the fourth images show volcanic eruptions with only smoke columns; the fifth image shows a volcano in which no eruption is happening; the other four images are random images with a city (the seventh one).

8.2.4   Proposed model

The binary classification task has been addressed, by implementing a classifier with two classes: a first class for the eruptions and a second class for the other scenarios.

The proposed model has been trained with satellite optical data acquired by Sentinel-2 and Landsat 4, 5, 6, and 8, appropriately processed to highlight the presence of a volcanic eruption.

The model can be divided into two parts: the convolutional sub-network and the dense sub-network. The first one is responsible for the feature extraction of

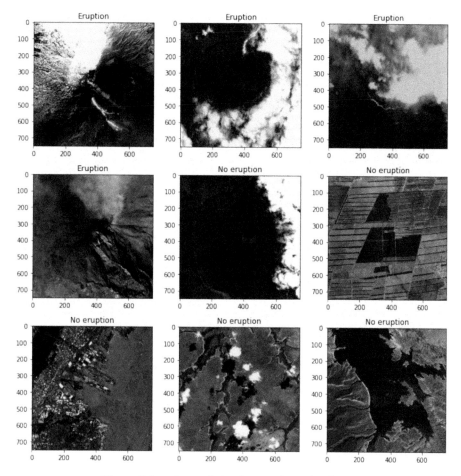

Figure 8.3   A set of nine images from the dataset

the images, the second one is responsible for the input classification. The first sub-network consists of seven convolutional layers, followed by the Rectified Linear Unit (ReLU) activation function, a pooling layer, and a batch normalization layer. Each convolutional layer has a stride value equal to (1,1) and a variable number of filters. Each max-pooling layer with kernel size and stride of (2,2) allows to halve the feature map.

This type of network is able, layer by layer, to extract information from the input data and to compress this information into sub-matrices or feature maps gradually. This is possible both because of the number of filters created with the network's profile and because at each level the feature map is halved. At the last level the feature map will have a very small size and a very high depth. It is representative of the level of abstraction that the network has reached.

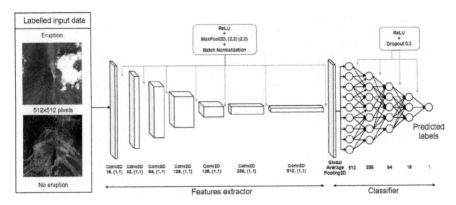

Figure 8.4   Network architecture

Network inputs are matrices/images of 512×512 pixels and a depth of 3 (RGB). The output is a matrix of depth equal to 512 (number of filters) and size 4×4 pixels.

To connect this first sub-network to the second sub-network, it is typically necessary to transform the feature map of the last layer into a one-dimensional (1D) vector. This can be done by a layer called flatten or by a layer called global average pooling. The substantial difference between the two is in the number of effective parameters that the network will have to optimize.

The second sub-network consists of five layers in which the first four are hidden. Each layer is followed by the ReLU activation function and a dropout layer. Also in this case the number of neurons is made to vary in decreasing way.

The input of this network will therefore be a vector, a rolled out feature map, while the output will be a digit and a probability. The first indicates the class to which the input belongs, while the second represents the probability of its validity.

The overall network architecture is shown in Figure 8.4.

The strength of this type of network lies in the correct sizing of the hyperparameters. In fact, the network described above is a version obtained after a series of optimization processes. Furthermore, the choice of using global averaging pooling as a flatten layer has made it possible to drastically reduce the parameters and therefore increase the dimensionality of the network without incurring in memory problems.

The model has been trained using the binary Cross-entropy as loss function. This type of function is commonly used for problem of classification with two classes. Cross-entropy loss, or log loss, measures the performance of a classification model whose output is a probability value between 0 and 1. Cross-entropy loss increases as the predicted probability diverges from the actual label. The binary cross-entropy formula is

$$H_p(q) = -\frac{1}{N}\sum_{i=1}^{N} y_i * \log(p(y_i)) + (1 - y_i) * \log(1 - p(y_i)) \qquad (8.2)$$

where y is the label (0 or 1) and $p(y)$ is the predicted probability of being 1.

8.2.4.1   Image loader

As the training dataset for no eruption is bigger than the training dataset for eruption, a different approach should be implemented. To obtain good results, the batches should contain an equal number of instances from the two classes. For this reason an arbitrary function called Image Loader from the Phi-Lab *ai4eo.preprocessing* library has been used.

This library allows you to define a much more efficient image loader than the already existing Keras version. Furthermore, it is possible to implement a data augmentator that allows to define further transformations.

The most powerful feature of this library is the one related to the balancing of the dataset. In fact, as anticipated in the previous section, an unbalanced dataset, with a number of examples of one class greater than the other, will lead the model to recognize only the dominant class.

The dataset created, by its nature, has a number of examples of volcanic eruption decidedly lower than the non-eruption class. Since reducing the number of examples can lead to the training of a weak network, it was necessary to balance the weak class through the oversampling technique.

8.2.5   Training

During the training step, for each epoch the error between the real output and the prediction is calculated both on the training dataset and on the validation dataset. The metric used for the error is the accuracy, the ratio of the number of correct predictions to the total number of input samples, which is given by

$$\text{accuracy} = \frac{\text{number of correct predictions}}{\text{total number of predictions}} \qquad (8.3)$$

It works well only if there is an equal number of samples belonging to both classes.

The model has been trained for 100 epochs, using the Adam optimizer. The model has a number of parameters to be optimized equal to 2,136,385 of which only 2272 cannot be optimized.

The training dataset is composed of 1215 examples, of which 334 are with eruptions and 818 are without eruptions. The validation dataset contains 75 "eruption" images and 94 "no eruption" images. Both datasets are subjected to the data augmentation and to the addition of noise to increase the robustness of the model.

During each time a batch of 16 images is sent to the model via the generator. The generator allows you to manage memory properly: this technique is particularly used when working with graphics processing units (GPUs).

Through this procedure the model employs, during the training phase, about 370 s to complete a training epoch. These values were obtained by training the model on the Google Colaboratory platform. Remember that this tool presents a GPU1 ×Tesla K80, having 2496 CUDA cores, computer 3.7, 12G (11.439GB Usable) GDDR5 VRAM, a CPU 1×single core hyper threaded, i.e (1 core, 2 threads) Xeon Processors @2.3 GHz (No Turbo Boost), 45 MB Cache, 12.6 GB of available RAM, and 320 GB of available disk. The trends of the training and the validation loss and accuracy functions are shown in Figure 8.5.

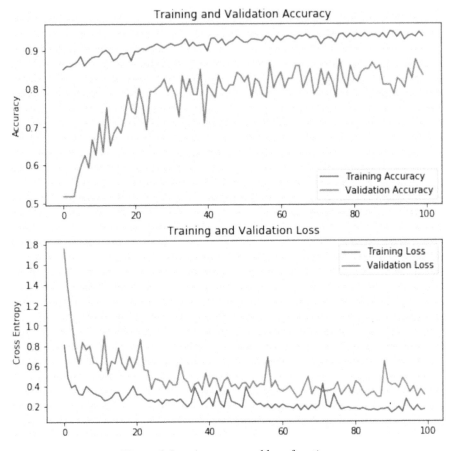

Figure 8.5   Accuracy and loss functions

After 100 epochs, the model has reached a loss value near 0.2 for the training and 0.4 for the validation, and an accuracy value next to 0.95 for the training and 0.85 for the validation. Furthermore, using the early stopping technique is possible to avoid overfitting and to save the model that has reached the highest validation accuracy value.

In Figure 8.6 the confusion matrix is shown.

Among 89 images without volcanic eruption 76 have been classified correctly and the other 13 are false positive, while among the 71 images of volcanic eruptions 11 have been misdetected and the other 60 have been classified correctly. Considering the different numbers of images in the two classes, the percentage of right and wrong classification is the same for both eruption and no eruption class. The 85% of images are classified correctly and the other 15% are classified as the opposite class.

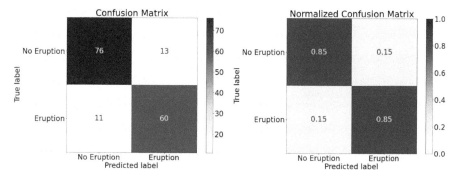

Figure 8.6   Confusion matrix

8.2.6   Test on test dataset

This section will present some results on tests performed on the test dataset. The tests are presented in three batches of nine images for each one. Above each image the real label is reported while in the tables a comparison between the real label and the predicted one is shown.

The predicted label is not represented as a binary number (0 or 1) but by a float number between 0 an 1: in this case the precision of the prediction can be evaluated. A low value, next to 0, indicates a low probability of eruption, so the image is classified as "no eruption"; a high value, next to 1, indicates a high probability of eruption, so the image is classified as "eruption"; a value next to 0.5 indicates a random prediction.

As can be seen from Table 8.3 and Figure 8.7, in this batch the model has achieved good performance apart from the first case in which a misdetection occurred. This case is particularly important because there is no evident eruption happening. This image has been labeled as eruption because it was acquired in a period when there was an eruption according to the catalog. All the other images have been classified correctly.

Table 8.3   Results on test dataset (1)

Image	Real label	Predicted label
Row 1–Column 1	1	0.13
Row 1–Column 2	1	0.99
Row 1–Column 3	1	0.99
Row 2–Column 1	0	0.03
Row 2–Column 2	1	0.99
Row 2–Column 3	1	0.99
Row 3–Column 1	0	0.0
Row 3–Column 2	1	0.99
Row 3–Column 3	0	0.0

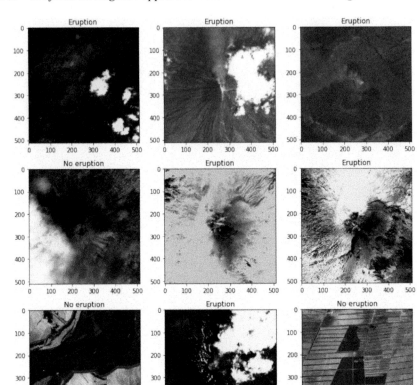

Figure 8.7   Test on test dataset (1)

Table 8.4   Results on test dataset (2)

Image	Real label	Predicted label
Row 1–Column 1	1	0.94
Row 1–Column 2	1	0.98
Row 1–Column 3	1	0.99
Row 2–Column 1	0	0.0001
Row 2–Column 2	1	0.99
Row 2–Column 3	1	0.99
Row 3 – Column 1	1	0.99
Row 3–Column 2	0	0.000002
Row 3–Column 3	0	0.02

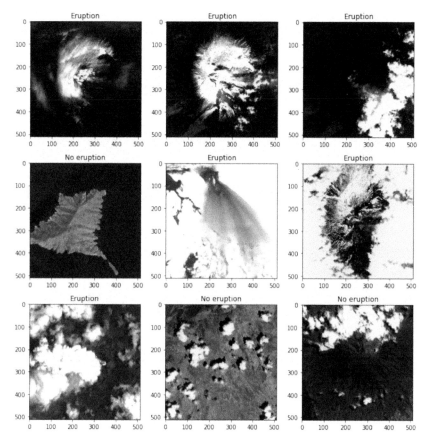

Figure 8.8   Test on test dataset (2)

In this case (see Table 8.4 and Figure 8.8), the model has achieved very high performances, in fact all the images have been classified according to the catalog.

In this case the model has not achieved very good results (see Table 8.5 and Figure 8.9). Two types of problems occurred. On image 5 a higher value has been predicted. Observing the type of image, to avoid this, other volcanoes' images with clouds could be added to the dataset. However, the problem is of low entity and the label is still correctly predicted. The biggest problem is on image 8. In this case a "no eruption" image is classified as "eruption" providing a false positive. This may be due to the crater shape and the presence of a fluid in it, perhaps identified as magma [2].

8.2.7   Future works

In order to improve the results, some ideas can be followed:

Table 8.5   Results on test dataset (3)

Image	Real label	Predicted label
Row 1–Column 1	0	0.0003
Row 1–Column 2	0	0.0001
Row 1–Column 3	0	0.00002
Row 2–Column 1	0	0.0
Row 2–Column 2	0	0.26
Row 2–Column 3	1	0.99
Row 3–Column 1	1	0.99
Row 3–Column 2	0	0.88
Row 3–Column 3	0	0.01

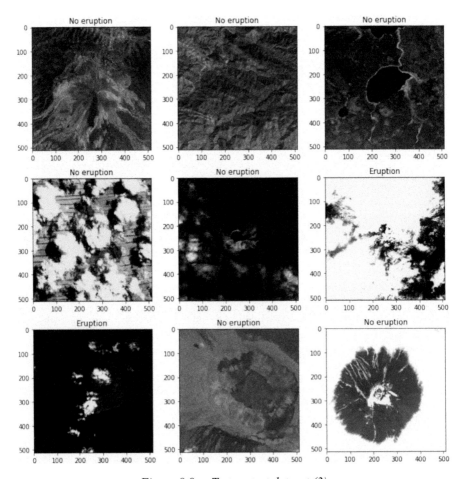

Figure 8.9   Test on test dataset (3)

- Use the infrared bands independently: the information will no longer be merged into a single RGB image but will remain divided into different bands and independent of each other.
- Increase the spatial dimension of the data to include more details in the scene: this involves an increase of the training time and a possible change in the network in order to make it learn better the new features and discern the ones essential for the detection.
- Increase the dataset with other optical data and with events not present in the catalog.
- Improve the model structure to achieve greater accuracy.

An important future step could be the use of Sentinel-5P data. Sentinel-5P, precursor of Sentinel-5, is the first Copernicus mission dedicated to monitoring our atmosphere. The satellite carries the state-of-the-art Tropomi instrument to map a multitude of trace gases such as nitrogen dioxide, ozone, formaldehyde, sulfur dioxide, methane, carbon monoxide, and aerosols. For this reason it could be used for monitoring the gas emissions during a volcanic eruption. The network could be changed to include as input both information on visual effects and atmospheric composition during a volcanic eruption to better detect this one. The use of this type of information can also make it feasible to predict an imminent eruption. An example of information acquired by Sentinel-5P during a volcanic eruption is shown in Figure 8.10.

8.3 Case study B: landslides binary classification

In human history, for reasons mainly related to the need of hiding from easy finds, housing settlements have been built in many hidden places, and often at the foot of mountains, since these sites have inspired a sense of security and protection from prying eyes. Ironically, these former locations turned out to be the worst and most dangerous places, because at a later stage, the descendants intervened heavily in changing the surrounding environment. Deforestation, rain, and other weather phenomena were the main triggers for many landslides. Different types of landslides can be identified and fitted into a systematic classification scheme, based on the type of movement of the sliding mass and the displaced materials, as explained later in this section. Unfortunately, sliding and falling masses often occur without clear warning signs, and the devastation they bring often results in serious infrastructure damages or destruction and severe losses of human lives [3–8].

Automated detection of landslides from satellite data can help reduce their harmful impact by offering early warning, immediate alert, and clear mapping of the affected area. In this context, spaceborne radar data have been widely used in landslide mapping research, thanks to their global coverage, high spatial and temporal resolution, and all-weather capability. In [9], for example, the authors use InSAR techniques to monitor and analyze the deformation that preceded and resulted from a landslide in Guanling, China. In [10], the authors report on a method for landslide

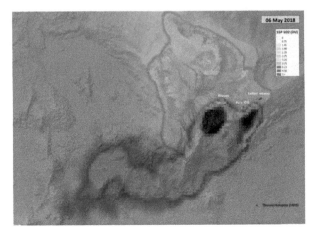

Figure 8.10   *Example of information acquired by Sentinel-5P during a volcanic*
eruption. Cit: https://earth.esa.int/web/sentinel/news/content/-/
asset_publisher/BZewkR1itkH2/content/copernicus-sentinel-
5p-spots-so2-emissions title: Copernicus Sentinel-5P spots SO₂
emissions. Description: On May 6, 2018, Sentinel-5P and VIIRS
(instrument on the NASA/NOAA satellite Suomi-NPP) detected
sulfur dioxide emissions and lava (hot spots) from the Kīlauea
volcano in Hawaii. Copyright: BIRA/IASB

characterization and monitoring, based on persistent scatterer interferometry (PSI), while in [11], the integration of PSI and GB-InSAR (ground-based synthetic aperture radar interferometry) is applied for mapping a landslide in Sicily, Italy. In general, InSAR has been resulted in a powerful technique for geohazard assessment and critical infrastructures monitoring, as discussed in [12], and the use of spaceborne image analysis has helped in land cover classification and evaluation of landslide damages [13, 14]. Yet, recently researchers worldwide have moved from the application of more traditional techniques, such as the creation of high-resolution digital elevation models (DEMs), multisensor images and data combination, InSAR and DInSAR techniques, to use machine learning (ML) techniques for geohazard assessment, and landslide susceptibility mapping. In the context of a broader use of spaceborne remote sensing, landslide susceptibility mapping can be produced with good results, through logistic regression and support vector machines [15–20], but these methods often rely on exact features of the geographical region, such as elevation, slope inclination and soil type, or extensive preprocessing of the images. The objective of this section is to present a method based on convolutional neural networks (CNNs) and freely downloaded satellite data, whose main strengths are: (1) the capability of the CNN to offer almost instantaneous response, after the training period is completed and (2) the capability to recognize a landslide in any context, thus producing outputs with a minimum amount of ancillary information and without the use of threshold-based algorithms. The CNN, presented in [21] and processing Sentinel-2 images, is analyzed by showing its ability to detect landslides with a good performance. Results in fact show that the deep learning approach can be used for this aim, but the task, being quite complex, requires a good construction of the dataset, a good preprocessing phase, and network design. Some issues were left open, such as the possibility to extend the algorithm to SAR data, which would allow to overcome the limits due to the cloud covering, when optical images are used; the validation of the proposed method on different types of landslides, in terms of size, by also taking into account small and medium-sized landslides, and in terms of their triggering causes. As regards this latter point, it is worth highlighting that different causes, acting alone or in combination, such as earthquakes, heavy rains, snow melting, and deforestation, can generate different landslides, and the identification of the landslide typology can help the visual interpretation of the images, as presented ahead in this section. A completely new CNN has been therefore developed and presented in this section, applied to Sentinel-1 images, and to landslides different in size and type (i.e. for triggering causes). To the best of the authors' knowledge, the application of CNNs to public satellite images of landslides is at the beginning of the exploration, as also evidenced by the paper [22], and there are no cases in the literature where CNNs have been applied for landslide detection and classification, based on the geological and geotechnical interpretation of the landslides.

8.3.1   Landslide classification

Landslides may be generated by different causes acting alone or in combination. Landslide triggers may be due to heavy rains (high intensity or moderate rain combined to longer duration), snow melting, river flood or significant change in water level (typical of dam reservoir), earthquakes and related phenomena, such as liquefaction. The further classification is therefore based both on the type of the failing material (rock, debris, soil) and the type of the sliding mass movements (falls, topples, slides spreads, flows), according to the criteria proposed by [23], and adopted by many agencies, such as the California Geological Survey (CGS) [24]. Reagrding landslide movements, falls are masses of soil or rock that dislodge from steep slopes and roll down. Topples move by the forward pivoting of a mass around an axis below the displaced mass. Slide spreads, commonly induced by liquefaction of material in an earthquake, move by horizontal extension and shear or tensile fractures, by displacing masses of material along one or more planes. In rotational sliding, the slide surface is curved and the mass rotates backward around an axis parallel to the slope. In translation sliding, the failure surface is more or less planar and the mass moves parallel to the ground surface. Flows mobilize as a deforming, viscous mass without a discrete failure plane. Moreover, more than one form of movement may occur during a failure, in which case the movement is classified as "complex" or "composite." The main possible material/movement combinations, commonly considered in preparing a landslide inventory map, are the following: earth flows, debris flows, debris slides, rock slides, and rock falls, described in more detail as follows [24]:

- Earth flows commonly occur on moderately steep slopes (slope gradients from 10% to as steep as 30%). Earth flow triggers are periods of prolonged rainfall. They are characterized by slow movements, in the millimeters or centimeters per day range, and may continue to move for a period of days to weeks after initiation.
- Debris flows may be triggered by intense rainfall following a period of less intense precipitation, or by rapid snow melt. High pore water pressures cause the soil or weathered rock to rapidly lose strength and flow downslope. Debris flows can move very rapidly, with rates ranging from meters per hour to meters per second and travel relatively long distances, making them very difficult to be detected promptly.
- Debris slides commonly occur on very steep slopes, as steep as 60–70%, usually in an area where the base of a slope is undercut by erosion. Debris slides move initially as shallow intact slabs of soil and vegetation, but break up after a short distance into falls and flows. Individual debris slides may move at rates ranging from meters per day to meters per minute.
- A rock slide is a landslide involving bedrock in which the rock that moves remains largely intact for at least a portion of the movement. Rock slides can range in size from small and thin to very large and thick. Rock slides commonly occur on relatively steep slopes in competent rocks. Slope gradients are commonly from 35% to as steep as 70%.
- A rock fall is a landslide where a mass of rock detaches from a steep slope by sliding, spreading or toppling and descends mainly through the air by falling, bouncing, or rolling. Intense rain or earthquakes may trigger this type of movement. Rock falls occur on steep slopes of hard, fractured rock. The scar left by a rock fall on the slope may be no more apparent than an area of rock that is less weathered than the surrounding rocks. Interested readers can refer to the CGS website for an even more effective and exhaustive summary [24].

In Figure 8.11, the main landslide types are shown together with two samples of Sentinel-2 and Sentinel-1 images, downloaded from the Google Earth Engine [25], by using as a reference the NASA Open Data Global Landslide Catalog (GLC) [26], where the date, the geographical position, the size, and other useful information are recorded. The rock fall and rock slide types are registered under the same ID in the catalog and for this reason the same images are associated to them in the figure. In Figure 8.11, optical images allow to see clearly the landslide shapes in almost all the cases, but they present the cloud cover issues, not present in the SAR images. In the case of the debris flow landslide, the analysis can still be done because the cloud does not cover the area of interest. Yet, in the case of the rock fall/slide, the cloud cover is more severe and the visual analysis can hardly allow to see the landslide. Moreover, it is worth saying that the same kind of images may appear different also because they may have been taken under different light conditions.

This analysis carried out on the geological characteristics of the landslides, and the visual inspection of related satellite optical and SAR images can allow to make some considerations. Whatever is the type of the landslide, there is a common

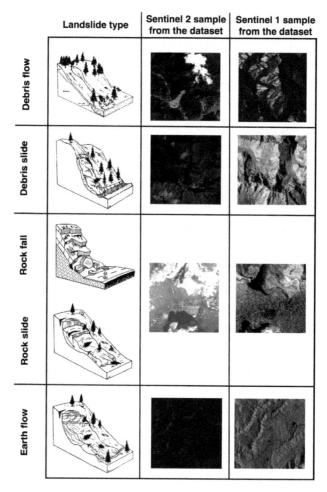

Landslide type	Sentinel 2 sample from the dataset	Sentinel 1 sample from the dataset

Figure 8.11 *Main landslides types with related Sentinel-1 and Sentinel-1 images (from top to down): debris flow at Xinmo (China), on June 24, 2017; debris slide at Piz Cengalo (Switzerland), on August 23, 2017; Rock fall at El Capitan in the Yosemite National Park, California (United States), on September 28, 2017; Earth flow at Morton Road, Dutch Flat, California (United States), on January 11, 2017*

aspect: a *wound*, a *scar* in the area of interest, with a modification of the surface and a variation of a certain entity in the involved ground mass or rocks. These considerations have convinced the authors that the application of methods based on feeding the neural networks with images collected before and after the event could work. Moreover, convolutional networks appear to offer more advantages if compared, for instance, with a simple pixel-based algorithm, as discussed in the next paragraph.

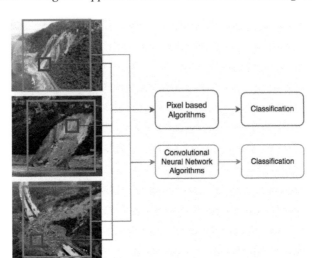

Figure 8.12   CNN s, pixel-based algorithms

The evident change in the pre- and post-event images have also inspired some of the future research activities, as explained in Section 8.3.7.

8.3.2   The CNN for landslides binary classification

As clearly evident from the graph shown in Figure 8.12, the classical pixel-based algorithms rely on a single pixel or on small groups of pixels at most. It is thus easy to understand that an algorithm of this type would report a landslide if it encounters in the image a pixel or groups of pixels that, during the training, are associated with a landslide (e.g. based on the color). Differently, the algorithms based on CNNs work on large portions of the images, by extracting particular features (e.g. the landslide shapes, which are distinct from the shapes of other natural phenomena) and producing more reliable results. In particular, the convolution is carried out between an input image and one or more matrices (of a smaller size), called kernels or filters. In the case of a two-dimensional (2D) image I as input, and a 2D kernel K, the kernels are varying at runtime and modified by the network itself, which adapts them in order to capture the most important features present in the images [27–30].

8.3.3   The method based on optical images

The method, proposed in [21], is based on the use of a CNN applied to optical Sentinel-2 images, and boundary information, to determine whether the landslide occurred. The network, described by weights and biases, and fed with the selected images as input, performs convolutions to produce the output. Random values for the initial weights and the negative log-likelihood function as loss function have been chosen, as given by

$$L(X) = -\sum_{x,y \in \chi} p(x) \cdot \log(y) + (1 - p(x)) \cdot \log(1 - y) \qquad (8.4)$$

with X being the training dataset of images (label pairs) and $p(x)$ the prediction (output) of the model. The backpropagation algorithm and the Adam optimizer[31, 32] were used in order to repeatedly update the weights in the direction that minimizes the loss function. It is worth saying that authors have analyzed relevant publications where different choices for the initial weights and updated versions of the Adam optimizer have been tested [33–36]. They are going to take into consideration these modifications in their future work. In Figure 8.13, the block scheme of the CNN proposed for the optical satellite data is shown.

Pairings of images before the landslide, as examples of no landslide images, have also been retrieved. The developer console of Google Earth Engine [25] has been used for preprocessing the images. Several operations were carried out during this stage: selection of images by cloud cover; random cropping of a single landslide image to produce smaller images guaranteed to contain the landslide; image rotations and flipping to increase the number of images available for training the network (data augmentation). This latter contributes significantly to make the model gain in accuracy. After the preprocessing stage, the Sentinel-2 images are fed into the CNN model, together with some historical weather data (rainfall, humidity, cloud cover). The CNN is built on eight learning layers. A single input consists of two images (one before and one after the event), each of size 512 by 512 pixels, and an extra dimension of size 5 for the Sentinel-2 bands. Bands 2, 3, 4, 8, and 12 are used because of the level of resolution and the independent information they provide, as it can be easily deduced from Figure 8.14.

This forms a $2 \times 512 \times 512 \times 5$ input, which is then processed using 3D convolutions. An example of each pair of processed images is shown in Figure 8.15, where Sentinel-2 data acquired before and after the landslide in Xinmo (China) are shown. In the second image the landslide is clearly detectable, even if a cloud is present, because it does not cover the area of interest. The acquisition date refers to the post-event image.

The model was trained on 20 different landslides, using a fivefold cross-validation method to corroborate on four landslides at a time. Cross-validation is a resampling procedure primarily used in applied ML to estimate the skills of an ML model on unseen data [38]. It is used to evaluate ML models on a limited data sample, and it estimates how the model is expected to perform in general, when used to make predictions on data not used during the training of the model (unseen data). It has a single parameter called k that refers to the number of groups that a given data sample is to be split into. Therefore, the procedure is often called k-fold cross-validation. When a specific value for k is chosen, it may be used in place of k in the reference to the model, such as $k = 10$ becomes 10-fold cross-validation. It is a popular method because it is simple to understand and because it generally results in a less biased or less optimistic estimate of the model skills than other methods, such as a simple train/test split. The general procedure is briefly described as follows:

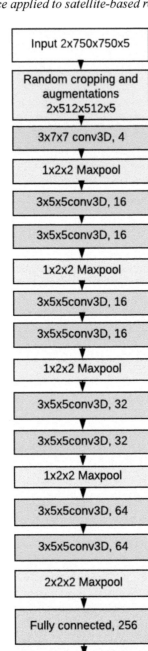

Figure 8.13   Block scheme of the proposed CNN for Sentinel-2 images

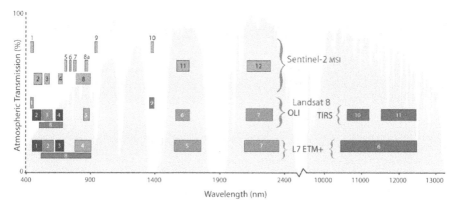

Figure 8.14   Sentinel-2 and Landsat bands [37]

1. Shuffle the dataset randomly.
2. Split the dataset into *k* groups.
3. For each unique group:
   (i)   take the group as a hold out or test dataset
   (ii) take the remaining groups as a training dataset
   (iii) fit a model on the training set and evaluate it on the test set
   (iv) retain the evaluation score and discard the model
5. Summarize the skill of the model using the sample of model evaluation scores.

Importantly, each observation in the data sample is assigned to an individual group and stays in that group for the duration of the procedure. This means that each sample is given the opportunity to be used in the hold out set 1 time and used to train the model $k - 1$ times [39]. For evaluating the accuracy of the network, the Keras balanced accuracy has been used as performance parameter (in this case, yes landslide and no landslide), which is calculated as the mean of the accuracy values from each sample class [40]. For example, a network that predicts every sample to be a landslide would achieve a 100% accuracy on landslides and 0% accuracy on

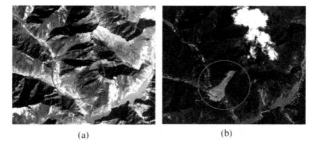

(a)   (b)

Figure 8.15   An example of Sentinel-2 images acquired before (a) and after (b) a landslide (Xinmo landslide, China, June 24, 2017)

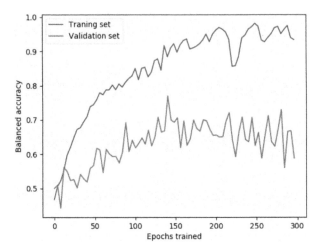

Figure 8.16   Training and evaluation accuracy

images without a landslide, resulting in a balanced accuracy of 50%, regardless of the relative frequencies of the classes. The network has been trained for 120 epochs with the training set. On the validation set it reaches an average balanced accuracy of 0.624 on detecting the landslides.

As shown in Figure 8.16, the accuracy values are quite volatile and significantly higher than those achieved while training the network. This is most likely due to the relatively small number of samples. Nevertheless, it is important to note that the proposed algorithm is able to pick up significant signal after being trained on just a small sample of optical imagery. This is remarkable considering that typical deep learning algorithms require at least a few hundred samples from each class [41]. A larger-scale approach is likely to achieve even better results. Given that the model was slightly overfitting, as revealed by the diverging training accuracy and valida-tion accuracy, the authors have worked on developing a new network to be applied to Sentinel-1 SAR data, which will be introduced in the next section. In addition, the authors will demonstrate how the new CNN works not only with very large and catastrophic landslides, but also with large, medium, and small-sized landslides. These objectives are in line with the future work highlighted in [21].

8.3.4   SAR data and creation of the dataset

The use of radar data in place of optical data presents several advantages. The most widely known is robustness to cloud cover and in general also to atmospheric conditions. The sensitivity of radar backscatter to the shape of the observed object at different scale also plays an important role. Detecting a landslide with optical data primarily relies on the change in spectral response induced by the change in the material at the outermost layers. Radar data are instead related to the geometry of the surface and can gather more pieces of information, including for instance the DEM, the slope of a land, the surface roughness, and the overall 3D shape of the observed area [42–45].

First of all, the authors have examined which kind of SAR data should be used to extract the right features for the detection purpose. Two types of SAR data from the satellite Sentinel-1 have been analyzed: the Level-1 Single Look Complex (SLC) product and the Level-1 Ground Range Detected (GRD) product.

8.3.4.1   GRD and SLC SAR data

SLC products consist of focused SAR data, geo-referenced using orbit and attitude data from the satellite and provided in zero-Doppler slant range geometry. Each image pixel is represented by a complex (I and Q) magnitude value and therefore it contains both amplitude and phase information. The products include a single look in each dimension using the full transmit signal bandwidth.

GRD products consist instead of focused SAR data that have been detected, multi-looked, and projected to the ground range, by using an Earth ellipsoid model. Phase information is lost. The resulting product has approximately square spatial resolution pixels and square pixel spacing with reduced speckle at the cost of a coarser spatial resolution [46]. The Google Earth Engine catalog contains only GRD data in dual polarization (VV and VH). For this reason in the proposed CNN to be used with Sentinel-1 images, the focus has been on GRD data, and in order to increase the information level of the final data, used to train the network, an RGB composite has been created. The following formulae (8.5a–c) have been applied:

$$R = VV \tag{8.5a}$$

$$G = VH \tag{8.5b}$$

$$B = \text{change}(VV, VH) \tag{8.5c}$$

where in equation 8.5c, the change function returns the change map between two images (see Algorithm 8.1) and it is based on a *virtual* normalized difference vegetation index (NDVI), which authors called *modified* NDVI, inspired by the NDVI calculated from optical data, but based on SAR information.

Algorithm 1: Change map algorithm

```
normalize(VV);
normalize(VH);
modifiedNDVI = (VV-VH)/(VV+VH);
for i in width do
    for j in height do
        if modifiedNDVI[i][j] ≥ threshold then
            |   changeMap = 0;
        else
            |   changeMap = 1;
        end
    end
end
```

Table 8.6   *A sample of landslides of interest*

Location	Event date	Size	Latitude	Longitude
Greenland	June 17, 2017	Catastrophic	71,53659933	−53,20874578
California	May 20, 2017	Very large	35,865628	−121,43238
China	June 24, 2017	Catastrophic	32,08087401	103,6656168
Indonesia	August 19, 2016	Very large	−6,311708	106,801076
Colombia	March25, 2017	Very large	1,15189804	−76,639923
Switzerland	August 23, 2017	Very large	46,29694	9,595744
United Kingdom	June 29, 2017	Very large	50,70838	−2,75802
California	September 29, 2017	Large	−119,63568	37,72978945
Switzerland	August 23, 2017	Very large	9,5957443	46,29694459
California	January 11, 2017	Medium	−120,8095385	32,20099693
...	...	...	...	...

Algorithm 8.1 and (8.5) allow to highlight the differences between the VV and VH polarization, but at the same time to increase the dimensionality of the data. In fact, aiming at a data fusion approach between optical data and SAR data, it is necessary to ensure that the inputs of the two networks proposed for optical and SAR data are consistent. The dimensionality in fact changes by doing so from $1\,000 \times 1\,000 \times 2$ (width $\times$ height $\times$ number of polarizations) to $1\,000 \times 1\,000 \times 3$ (width $\times$ height $\times$ (number of polarization + change function)).

As mentioned before, the data were downloaded using Google Earth Engine and the NASA Open Data GLC as a reference, with all the information about the date, the geographical position, the size of each landslide. This catalog indicates the position accuracy for each landslide (i.e. perfect accuracy, unknown accuracy, and greater then 1 km accuracy).

For the CNN applied to SAR data, landslides of different sizes have been taken into consideration (see Table 8.6), with respect to the case of the CNN for optical data, where only landslides with very large and catastrophic dimensions were taken. In any case, for the proposed CNN, it was decided to retain only landslides that were larger than the "small" class. The extension of the algorithm to this particular dimension will be analyzed in future studies. Furthermore, those landslides with a positioning error greater than 5 km and with an event date outside the Sentinel-1 mission start range have been discarded. At the end, the selected images, after and before each event, were those centered on geographical coordinates of the landslide and with a size of 1000×1000 pixels. Figure 8.17 shows the landslides chosen for the analysis carried out on the proposed CNN in the case of SAR data.

Due to position and date constraints, the final dataset contains more or less 400 images with a landslide and 400 images without landslides. An example of SAR data with a landslide, acquired in VV polarization, is shown in Figure 8.18. The date and the region are the same as those in Figure 8.15. As it can be seen from Figure 8.18, the terrain absorbs and reflects differently the electromagnetic wave transmitted by the SAR antenna, if portions of the land affected by the landslide are

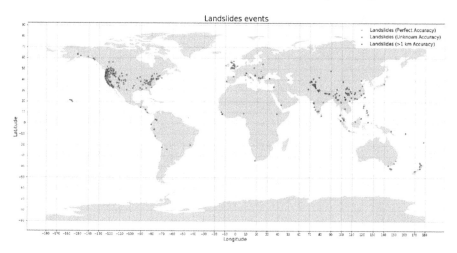

Figure 8.17   Plot of the landslides under analysis (NASA Open Data GLC [26])

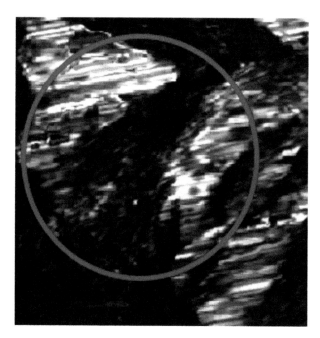

Figure 8.18   Sentinel-1 SAR image, acquired in VV polarization, of the same landslide of Figure 8.15 (zoom) (Xinmo landslide, China, June 24, 2017)

considered. The part of the image with the mountain marked by "streaks of white" is very different from the almost "black" part. This means that there has been a sudden change in reflectivity of that portion of land, such change being an effect of the landslide occurred.

An important objective, representing a starting point for using the algorithm, has been the creation of a dataset of suitable Sentinel-1 images. The relevant data were downloaded from the Google Earth Engine catalog through a tool developed by the authors and published on GitHub, the *Sentinel Data Downloader Tool* [1, 47].

8.3.5   The proposed CNN for SAR data

The model developed to be used with SAR data is a CNN, which performs a set of differentiable mathematical operations on the input values, described by a set of weights, to produce an output. Initially the weights are set to random values and then the network is trained to minimize loss function. For the proposed application the chosen loss function is the classic binary cross-entropy, mostly used in detection problems:

$$H_p(q) = -\frac{1}{N}\sum_{i=1}^{N} y_i \log(p(y_i)) + (1 - y_i)\log(1 - p(y_i)) \qquad (8.6)$$

where y is the label (0 for no landslide and 1 for landslide) and $p(y)$ is the associated probability that the input contains a landslide for all N inputs. The backpropagation algorithm and the Adam optimizer with a learning rate of 0.00002 were used in this case [31].

As can be seen from Figure 8.19, the proposed network architecture is composed of two main sub-networks. The first one is composed of eight convolutional layers and the second one is composed of five fully connected layers. Each layer of the convolutional mesh changes the dimensions of the input; in fact, this is increasingly compressed in height and width and expanded in depth. Through this mechanism the network, by applying different filters or kernels, is able to extract the features that are stored in the multiple dimensions of the input data. For example,

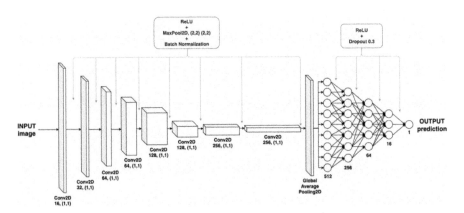

Figure 8.19   Proposed CNN architecture for SAR data

after the first convolutional layer, the data size will be shrunk by a factor of 2, to 500 × 500, due to the max-pool with parameter (2 × 2), on the first two dimensions; the depth instead extends from 3 to 16, due to the size of the convolutional layer being 16. Downstream of the convolutional layers, there is a global averaging layer which essentially transforms the output data to the last convolutional layer from matrix to vector. Then we move from a 125 × 125 × 256 matrix to a 1D vector with a length equal to 125 × 125 × 256 elements. This step is necessary to connect the section of the convolutional network to the dense one. The dense network, on the other hand, is made up of several fully connected layers. It deals with using the vector leaving the global averaging layer as input to decide whether or not there is a landslide in the input image.

In order to train the network in a faster and better way, the Colab platform of Google [48] was used. This platform allows to train models through the GPUs, free of charge. GPUs allow parallel calculation and are able to speed up the training process of a CNN by five times compared to a classic CPU [49]. To further speed up the training process, an image generator was also used. An image generator is necessary when the computational resources are scarce and the number of data items is moderate, in fact it is often associated with data augmentation. Considering that each image is a 1000 × 1000 × 3 matrix, it is really difficult for a computer to store a dataset of *N* images (*N* × 1000 × 1000 × 3 matrix). For this reason, the choice has been to start only with 16 images, to be loaded into the memory at a time, and as soon as the life cycle of the first 16 images ends, the generator will overwrite them with another 16. In this way, for each training instance, only 16 images are loaded into the memory. The following procedure describes the operations of an image generator.

1. Preprocessing stage:
   (i) Load and store the images full paths (first the images with landslides and then the images without landslides).
   (ii) Create a vector with the labels (1 landslide, 0 no landslides).
   (iii) Shuffle the paths and labels vectors with the same seed.
4. Training stage:
   (i) elect a random subset of paths and labels.
   (ii) Read the images by using the paths.
   (iii) Process the images.
   (iv) Yield the batch (images and labels).

In summary, the generator loads the path of all the images from the dataset into memory. During the path loading process, labels are assigned based on the folder in which the images are saved. So the images in the "no landslide" folder will have a "0" label and the images in the "landslide" folder will have a "1" label. After loading the paths and labels in vectors, a random shuffle is applied to both with the same seed (in order to respect the correspondence of the label image). During the training the generator extracts from the vectors of paths and labels a number of images dictated by the batch size. Through the paths sub-vector, the generator loads the image through specific functions, applies a normalization algorithm (to force the range of values to be between 0 and 1), applies data augmentation (rotation, flip, zoom, stretch, etc.), and returns (technically yield) two sub-vectors, one containing the images and the other the corresponding labels. By using the proposed version of the image generator, it is possible to use satellite images stored in a *.tif* format, which preserves the original pixel values in the images, without losing information. Other formats, such as *.png*, result into a scaling operation of the pixel values or into a normalization after censoring of outliers. To avoid this, the *.tif* format is preferred, and in this way it is possible to create a dataset whose values have not been manipulated and that has thus maintained its original range unchanged.

The dataset is made up of two sets: the training dataset and the validation dataset. Through the validation set at the end of each training epoch, the model is tested to verify the quality of the classification. The training set is composed of more than 400 images with landslides and more than 400 images without landslides. The validation set is created by splitting the training dataset (randomly taking a small percentage of images from the training set). The original number of available good images is in fact lower. There are many reasons because this happens. Given the position of the landslides, the data are often not usable or the acquisitions are in a period too distant from that of the event. Furthermore, the catalog contains many samples of small-sized landslides that are difficult to identify. It is worth emphasizing that the proposed algorithm has been tested on large and medium landslides, but one of the objectives is to test it also on smaller landslides.

8.3.6   Results

It is evident the complexity of the model: the total number of parameters to be optimized was 1,451,249, of which 1,449,361 were trainable and 1800 were not trainable. The model has been trained for 50 epochs and, for each epoch, the entire dataset, with a batch of 16 samples, was fed into the model, at each training step. Moreover, as already specified, the binary cross-entropy was chosen as loss function and the Keras accuracy as evaluation metric, given in this case by the formula:

$$\text{accuracy} = \text{mean}\left(\text{equal}\left(y_{\text{true}}, \text{round}\left(y_{\text{prediction}}\right)\right)\right) \tag{8.7}$$

where the function *equal* is the element-wise equality between the two tensors. Literature suggests that 0.5 is the correct threshold for the Keras accuracy to distinguish between classes [50].

As the reader can see from Figure 8.20, the values for training and validation accuracy tend to oscillate around 0.8 in both cases, but with greater fluctuations in the latter. In general, the model will not overfit, making it clear that the choice of the network, its size, and hyperparameters has been made correctly; moreover, it is performing better than the previous network created for the optical data.

As regards to the training and validation loss values, it is noticed that in the first case widely fluctuating values are reported, and they occur especially in correspondence with a big variation of the validation accuracy. This can be traced to a particular combination of images in a batch, for this reason it is important to carry out the maintenance of the dataset which we will discuss later. However, these two

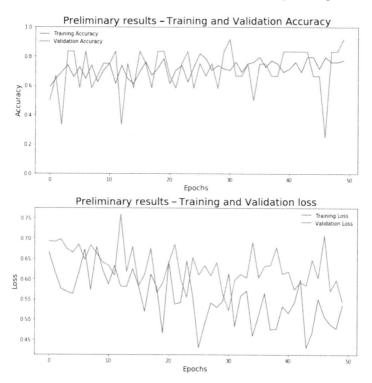

Figure 8.20   Training and validation accuracy (top). Training and validation loss (bottom).

graphs do not actually show all the potential of the network. At this aim, it is often used also for the confusion matrix.

We ran three different tests and calculated the related confusion matrices. In the first test, whose results are shown in Figure 8.21, it can be noted that the network is able to discriminate well the images with or without a landslide. However, it shows

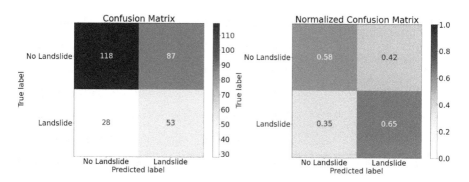

Figure 8.21   Confusion matrix network 1

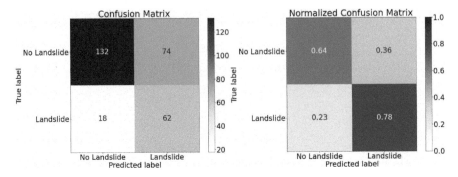

Figure 8.22   Confusion matrix network 2

a high percentage of false positives and a low percentage of false negatives, which represent the most dangerous situation. In general, false positives are preferable over false negatives. What has been done, it was to "tune" the network, by changing the hyperparameters and then retraining the network to reduce the number of false negatives, with a consequent increase in correct cases and false positives. However, the ideal goal would be no false positives/negatives and an equal ability to distinguish between landslide and non-landslide.

In a second test, a different selection of the network parameters has been implemented, through a long process of fine-tuning and modifications to the network (structural changes of the network, changes of the hyperparameters, and subsequent retraining). The final result, as it can be seen from Figure 8.22, was a better balance of the classes. Obtained results are better since both false positives and false negatives decrease and at the same time the correct decision classes significantly increase.

In the same way it can be ensured that it is possible to get an almost perfect balance between the classes, as it can be seen from Figure 8.23. In this third case, the compromise of having a greater number of false positives/negatives is accepted in exchange for a greater number of correct decisions. So it is understood that the

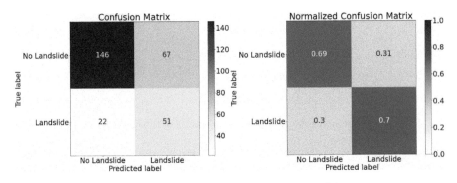

Figure 8.23   Confusion matrix network 3

Table 8.7   Results of each test sorted by relevance

Test number	Correct classification (no landslide) (%)	Correct classification (landslide) (%)	False positive (%)	False negative (%)
2	0.64	0.78	0.36	0.23
3	0.69	0.70	0.31	0.30
1	0.58	0.65	0.42	0.35

tuning of the network is a fundamental part for the creation of a network capable of approaching as much as possible an optimal decision.

All three cases show that the proposed network and the data used are suitable for this type of task, and at the same time the results show that there is still room for action, since it is likely that a further optimization of the network and its hyperparameters might result into a further reduction of false positives and false negatives.

As it can be seen from Table 8.7, the network designed under test 2 condition is the one that has the highest rates of correct decision, with greater strength for the detection of landslides and the lowest value of false negatives and for these reasons it is the first in the ranking. It follows the network designed under test number 3 which presents, unlike the previous one, a higher value of false negatives. Finally there is a network designed under test number 1, which actually has a low percentage of correct decision and high false negative/positive values.

For how the network is structured and how the dataset was created, the authors have shown that the use of convolutional networks, applied to SAR data, leads to good results. The structure of the available dataset, in which the images with landslides are separated by the images without landslides, allows to evaluate the network autonomously and without the need of comparisons with other results. The NASA Catalog in fact can be reviewed as a *ground truth* for validation. As previously mentioned , there are not models realized until now equal to those proposed by the authors. In addition, the ultimate objective is to combine optical and SAR data through fusion algorithms, in order to obtain a further boost in the accuracy of the model. Neural network classifiers are able to provide a good integration of different kinds of data, as highlighted in [51–53]. In fact, different data sources can be combined, because of their non-parametric approach, without the need to know the probabilistic distribution of the data. It is clear, however, that the task remains quite complex and initial insights will be discussed in the next section.

8.3.7   Future developments

The authors are currently working on reporting all their research in a book, which has been already accepted for publication [54]. Research activities are still in progress, and this section wishes to highlight some of them. As already underlined in this section, an Adam optimizer has been employed in order to repeatedly update the weights in the direction that minimizes the selected loss function. In the future the

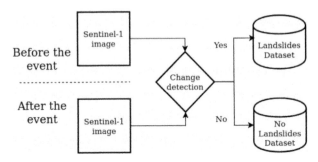

Figure 8.24   Change detection for the dataset creation

idea is to test different choices for the initial weights to be settled in the network and an updated version of the Adam optimizer as presented in [33–36].

8.3.7.1   Dataset maintenance

To ensure correct creation of the dataset, the authors, inspired by [55, 56], are developing a processing chain composed of change detection algorithms that can validate the presence of a landslide within an image [57, 58]. This step can enhance the detection of landslides in a SAR image, by ensuring that the dataset is correctly labeled. A diagram summarizing this part of the processing chain is shown in Figure 8.24.

8.3.7.2   RGB composite with SLC data

Another variant to be made in the future will concern the choice of the type of SAR data. Instead of using only GRD data, where phase information is missing, the idea is to use also SLC data. In fact, the SLC data are those necessary for the interferometric analysis, and to obtain coherence maps, DEMs and other pieces of information are useful to highlight the presence of a landslide. For this reason, authors are working on the creation of a new RGB composite, created from the coherence map (pre- and post-event) and from the combination of polarizations VV and VH.

To highlight the presence of a landslide, it is possible to increase the information level of the input images to be sent to the convolutional network. However, the processing chain proposed by the authors leads to a greater effort in the dataset creation phase and in the preprocessing phase. In fact, for the generation of the coherence map, it is essential to use the full-size SLC data, and this task has a high level of resource consumption, especially if the analysis is done hundreds of times, for applications related to deep learning, with a sufficiently large dataset.

For this reason the authors propose the use of the Open SAR Toolkit [59], a tool developed by Andreas Vollrath, a researcher of the European Space Research Institute. This tool automates Sentinel-1 data processing, it is written in Python and can be used in the Jupyter Notebook.

The idea proposed by the authors is to use a pair of Sentinel-1 images acquired with double polarization (VV–VH) to obtain a couple of coherence maps. Downstream of this process, the VV images and the corresponding coherence map

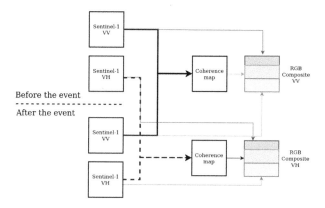

Figure 8.25   RGB composite summary scheme

are combined to obtain an RGB image in which the landslide is highlighted [60]. A summary scheme of this process is shown in Figure 8.25.

8.4   Case study C: crops-type classification

In this case study we present a way to use convolution neural networks, and benefit from their advantages, in case of a limited training set, which is common situation in Earth observations (EO). We show it on an example of crop-type classification (CTC) based on images acquired by drones.

8.4.1   CTC problem

CTC in EO domain aims at identifications of different crop types from a single or multi-temporal images. It can be considered as a subproblem of land cover classification for areas used to plant crops. Thus, it assumes the availability of crop mask (also called as crop extent), which differentiate crop land from other types of land cover, e.g., forest, water bodies, or artificial surfaces. CTC is vital for land monitoring and agricultural policies, especially nowadays when food security becomes an important issue [61].

Various methods have been developed so far for accurate mapping of crop types, including unsupervised and supervised methods. Recently, also solutions based on CNNs' architectures have been introduced [62, 63]. On the one hand, they bring a set of advantages over other approaches, including better performance. On the other hand, they form an important requirement on a large number of representative samples needed to train reliable model. In order to mitigate this problem, a semiautomatic active learning strategy can be applied [63]. This strategy is based on iterative process of extending a training set with most helpful unlabeled samples. In this paper we summarize our method on an automatic method for mitigating the problem of not sufficient number/representativeness of data for CNNs [64]. It is

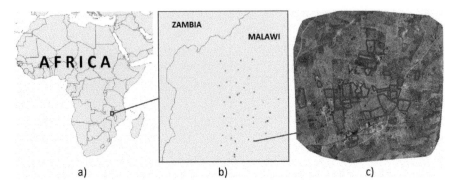

*Figure 8.26   Localizations of the orthomosaics (a–b) and selected orthomosaic
with perimeters of reference fields (c)*

based on using the networks trained first on computer vision datasets and then on
drone images depicting crop fields.

8.4.2   Dataset

A dataset used in this study was acquired in May 2018 in Kasungu district of Malawi.
It includes both (1) drone images and (2) reference information on crop types collected in a field campaign. Drone data consist of 40 orthomosaics acquired using a
CMOS camera with RGB bands. The ground resolution of a pixel is about 4 cm,
while each of the orthomosaics covers an area of about 1 km². Reference information is in a form of geo-localized polygons with annotations about crop type they
contain. It was prepared by enumerators sent to fields located in the areas depicted
by drones. The polygons were checked next and corrected by visual inspections
done by experienced interpreters. Localizations of orthomosaics and sample data are
presented in Figure 8.26.

   Five major crop-type classes are present in the dataset: maize, tobacco, groundnuts, cassava, and sweet potatoes. The rest of the crop land fields (containing minority crops, fallows, and harvested crops) were included in one class called "other."
Statistics on reference fields can be found in Table 8.8. Analyzing the statistics it is

Table 8.8   Statistics on reference fields

Crop-type class	No. of fields	Area (km²)
Cassava	32	0.123
Groundnuts	78	0.127
Maize	216	1.068
Sweet potatoes	37	0.025
Tobacco	61	0.142
Other	1214	7.389

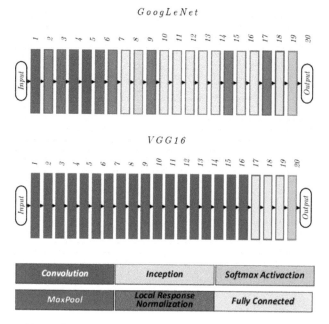

Figure 8.27   Architectures of CNNs used in the study

worth noting that the dataset is strongly unbalanced, "other" and "maize" classes are dominant in comparison to the rest of the classes.

8.4.3   Classification using CNNs and transfer learning

In order to classify drone images using the datasets described in the previous sub-section, we employed a patch-based approach. We used squared samples extracted from drone orthomosaics, each sample in size of 224×224×3 pixels according to the requirements of the networks we used later for classifications. The number of extracted samples was similar for each class (about 1000) to force classifier to learn on class features rather than on unbalanced distribution. The locations of samples were automatically randomly selected within the crop fields assuring that a polygon representing a field covers majority of a sample. Two popular CNN architectures were used to perform classification: GoogLeNet and VGG16. GoogLeNet [65] contains 22 layers with the considerably lower number of parameters (6.8 million) than other common networks used in computer vision. The core features of GoogLeNet are the inception modules, which perform convolution and max-pooling in parallel. VGG16 [66] contains 16 layers with 138 million learnable parameters. It consists of 13 convolutional layers and 3 fully connected layers. Architectures of both networks are compared in Figure 8.27.

The method used to train the networks was stochastic gradient descent with momentum with constant learning rate equals to 100. We used batch size of 100 samples and the network was trained for 20 epochs, which assured stabilization of learning results after fast improvements during firsts iterations. For both architectures trained with initial random weights, the achieved results were close to random due to limited representativeness of data in the training set. Additional randomly selected samples from the dataset cannot improve the results as they do not increase the representativeness. Representativeness could be improved only by extending the dataset with more crop fields, especially for minority classes. However, in this study, it was not possible as in many other cases of EO reference datasets. In order to overcome the problem of limited representativeness of training samples, we applied a method called transfer learning (TL) [67]. TL uses knowledge learned on images from source task to improve models learned in a target task. In this study, the target task was crop-type mapping on drone images. The source task should be different, but close to the target one. Thus, as a source task we selected computer vision object recognition due to the fact that for this problem many models trained on huge datasets are available. In particular, we used GoogLeNet and VGG16 networks with the weights learned on ImageNet dataset [68]. The original weights learned on computer vision dataset were used as initial ones for a training process (called fine-tuning in this context) conducted in the same way and with the same parameters as described above.

8.4.4   Results

Classification results were assessed using independent test set maintaining the real distribution of the classes. Results achieved for a TL approach were much better than random results obtained for the same CNNs trained from scratch. The overall accuracy obtained for GoogLeNet was close to 78%, while for VGG16 achieved almost 83%. The confusion matrices for both architectures are presented in Figure 8.28.

Among different classes, maize, groundnuts, and "other" are well recognized for both architectures. VGG16 were able to achieve good performance also for sweet potatoes. In both cases cassava and tobacco were the most problematic classes.

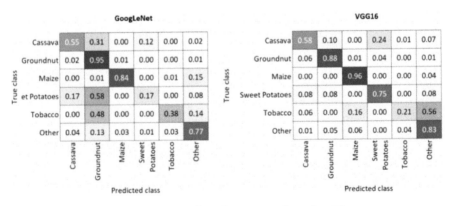

Figure 8.28   Classification results using TL

8.4.5   Future works

The case study demonstrates the advantage of TL on exemplary dataset. However, to obtain more general conclusions, it will be worth testing the methodology on other datasets, possibly also for other EO problems. Recently some initiatives to create huge annotated EO datasets have been started. A noticeable example is BigEarthNet project [69], which currently consists of more than 590,000 image patches annotated with labels taken from the CORINE Land Cover map of 2018. It is highly probable that the TL using CNNs pre-trained on this dataset could improve crop-type mapping results even more than computer vision-based, as in this case source and target tasks are much closer. However, it will be worth estimating how big could this potential further improvement be.

8.5   Case study D: segmentation

The concept of object segmentation, which dates back to the 1970s, even before the advent of the software that made it popular, is one of the basic projects on which remote sensing image analysis is built. In an interesting review [70] on object-based image analysis (OBIA), the history of the segmentation is reconstructed through the numerous articles reported, how we passed from an approach to classification based on pixels to an object-based approach, and its cruciality in OBIA.

Traditionally, the classification operates on individual pixels, using only the reflectance information contained in the image pixels to analyze and classify remote sensing images, ignoring spatial information or groups of pixels that should be considered together as an object [71]. In heterogeneous areas, such as urban landscapes, this approach has very limited applications due to the very similar spectral characteristics among different land cover types and high spectral variation within the same land cover type [72]. Myint *et al.* [73] demonstrated that traditional pixel-based approaches were not very effective in identifying urban land cover classes by running different tests on the entire QuickBird image.

With an object-based approach, by the segmentation process, the image is divided into segments, called objects, grouping pixels which have similar characteristics as brightness, color, and texture. This process decreases the variance within the same land cover type by averaging the pixels within the objects, which avoid the significant "salt-and-pepper effect," usually present in pixel-based classification. In addition to information related to each object (e.g. spectral signature, shape, size, etc.), information related to interrelationships among the objects can also be taken into consideration to identify the object.

The segmentation approaches can be grouped into three categories [74]:

- *Point based* – This approach searches and combines homogeneous elements that exhibit an equal or at least similar signal or characteristic value than a global threshold.
- *Edge based* – Margins are identified within the image through filters and an algorithm generates an outline that creates the object.

- *Region based* – Segments are found by growing the region (from bottom to top, i.e. starting from a seed pixel) or by dividing the region (from top to bottom, i.e. starting from the whole scene).

8.5.1   Segmentation tool and algorithms

The most used commercial software for image analysis in both academic and industrial fields is eCognition Developer, a product developed by Trimble [75]. eCognition is a powerful object-based development environment that allows to semiautomatically extract information from remote sensing data. It is used for the most common remote sensing tasks such as vegetation mapping [76–78], feature extraction [79, 80], and building extraction [81–83].

The segmentation algorithms offered by eCognition are as follows:

- *Chessboard segmentation* – divides the image into square objects with a user-defined size.
- *Quadtree segmentation* – creates squares of different sizes; very homogeneous regions produce typically larger squares than heterogeneous regions.
- *Contrast split segmentation* – segments the scene into a dark and bright image objects based on a threshold value that maximizes the contrast between them.
- *Spectral difference segmentation* – joins adjacent image objects if the difference between the intensities of their average level is less than the value given by the maximum spectral difference.
- *Multiresolution segmentation*(MRS) – groups areas of values of similar pixels into objects. Consequently, homogeneous areas result in larger objects, heterogeneous areas in smaller ones.

For technical details of each segmentation algorithms, readers can refer to the user guide [84]. Practical examples about MRS and *contrast split segmentation* will be shown in the next subsection.

8.5.2   Experimental analysis and results

The following examples will provide an overview of the steps to perform *contrast split* and MRS using eCognition software.

8.5.2.1   Contrast split segmentation

The *contrast split segmentation* is designed to segment images with significant contrasts. The purpose of this procedure is to divide luminous and dark objects through a threshold that maximizes the contrast of luminous objects (consisting of pixel values above the fixed threshold) and dark objects (consisting of pixel values below the threshold).

For this test, LiDAR's data, digital surface models (DSMs) and digital terrain models (DTMs), acquired in 2011 and retrieved from datasets of the extraordinary

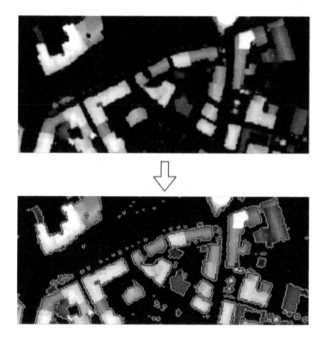

Figure 8.29   Contrast split segmentation on NDSM layer

plan of remote sensing [85], are used. Image objects using the *contrast split segmentation* are created through the following steps:

- Step 1. The normalized DSM (NDSM) raster was generated by subtracting the DTM from the DSM, to eliminate the effects of the topography (with respect to the terrain surface), using *NDSM layer calculation*.
- Step 2. *Contrast split segmentation* was performed on NDSM. The algorithm parameters, minimum threshold and maximum threshold (the minimum and maximum value of the range to use for contrast split segmentation), are defined as 182.88 and 274.32 cm, respectively.

In the test, a *contrast split segmentation* algorithm was used to separate tall objects from short objects based on NDSM (Figure 8.29). The results of this segmentation could be used in order to identify buildings or trees [83, 86].

8.5.2.2   Multiresolution segmentation

In the second example, MRS was performed on the WorldView-2 (WV-2) images, downloaded from the DigitalGlobe [75] image archive, as a standard Ortho-Readyproduct, projected on a plane with a Universal Transverse of Mercator projection and a World Geodetic System1984 datum.

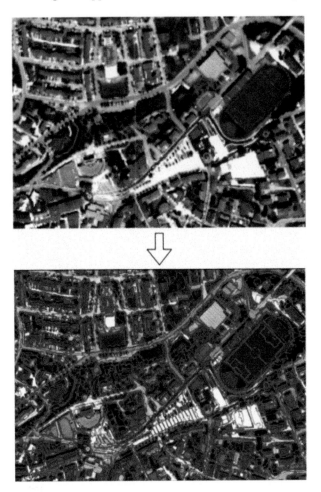

Figure 8.30   Multiresolution segmentation on WV-2 images layer

The WV-2 sensor provides a high-resolution panchromatic band and eight multispectral bands: four standard colors (red, green, blue, and near-infrared) and four new bands (coastal, yellow, red edge, and near-infrared).

The MRS algorithm has been performed using all multispectral bands of WV-2.

The most important parameters to be set in this algorithm are scale, compactness, and shape. The scale parameter (SP) defines the maximum degree of homogeneity of the objects and determines their size; higher values allow for more heterogeneity and produce larger segments. The compactness parameter defines whether the objects should be smoother or more compact. The shape defines the influence of color (spectral value) and shapes on the formation of segments [87].

In this example, the scale was set as 70, the compactness was set as 0.5, and the shape was assigned with a weight of 0.3.

The result of the segmentation is shown in Figure 8.30.

Segmentation creates objects that must then be classified. Therefore, the choice of segmentation parameters has a strong impact on the classification. The correct setting of parameters and the choice of thresholds are of interest to many researchers.

Polygons created using eCognition software, as seen, through segmentation can have different scaling factors. A smaller scale will correspond to smaller polygons because minor must result in the intra-polygon spectral variability and vice versa the larger the scale the greater the dimensions of the polygons [88]. Many studies used the trial-and-error approach to determine the SPs' segmentation values [73, 81, 89]. However, new automatic methods are proposed in the past years. Drăguţ *et al.* [90] introduced a fully automated methodology for the selection of SPs to perform MRS at three distinct scales, which relies on the potential of the local variance to detect scale transitions in geospatial data. Some techniques incorporate locally variable SPs into a region merging process, to segment out differently sized geo-objects in one segmentation result [91].

8.5.3 Integration of deep learning and object-based analysis

Feature extraction and selection in an OBIA approach need much supervision from expert knowledge to produce satisfactory results. It could be complicated in complex areas and time-consuming. For this reason, in the past years, some studies focused on the use of deep learning in combination with the object-based method for feature self-learning and automatic classification of complex urban. Zhao *et al.* [92] used MRS to generate image objects with precise edges; with the shape constraint of the image objects, deep features are combined with objects' features on the pixel level and used for image classification. In [93] the authors classify the LISS IV satellite images combining deep features and texture features obtained from OBIA method for accurate high-resolution image interpretation.

An object-based CNN under the OBIA framework using geometric character-izations was realized to guide the choice of sizes and locations of image patches; within-object and between-object information learnt by the OCNN model was also used to represent the spatial and hierarchical relationships [94].

As demonstrated in the reported studies, the integration of deep learning and OBIA significantly reduces the complexity of the feature set and also improves the classifier performance when compared to earlier approaches. It also avoids a waste of time of the analysis and could be really useful for many remote sensing application.

8.6 Conclusions

In conclusion, in this chapter we presented different case studies for image classifi-cation, target detection, and objects segmentation. The next chapters will introduce the reader to filtering problems and to generative models.

References

[1]   Sebastianelli A., Rosso M.P.D., Ullo S.L. Automatic dataset builder for machine learning applications to satellite imagery. Available from https://arxiv.org/abs/2008.01578 [Accessed September 2020].

[2]   Rosso M.P.D. 'Convolutional neural networks for the detection of volcanic eruptions. towards a new generation of satellites with artificial intelligence on board'. Thesis; 2019.

[3]   Althuwaynee O.F., Pradhan B., Lee S. 'Application of an evidential belief function model in landslide susceptibility mapping'. *Computers & Geosciences*. 2012;**44**(8):120–35.

[4]   Guzzetti F., Cipolla F., Lollic O., *et al. Information system on historical landslides and floods in Italy. Urban Hazards Forum*. CUNY, New York: John Jay College; 2002.

[5]   BGS. *Landslides around the world [online]*. 2019. Available from http://www.bgs.ac.uk/discoveringGeology/hazards/landslides/world.html.

[6]   Corominas J., Moya J., Ledesma A., *et al*. 'Prediction of ground displacements and velocities from groundwater level changes at the Vallcebre landslide (Eastern Pyrenees, Spain)'. *Landslides*. 2005;**2**(2):83–96.

[7]   Ferlisi S., Cascini L., Corominas J., *et al*. 'Rockfall risk assessment to persons travelling in vehicles along a road: the case study of the Amalfi coastal road (southern Italy)'. *Natural Hazards*. 2012;**62**(2):691–721.

[8]   Spizzichino D., Margottini C., Trigila A., *et al. Mapping The Impacts Of Natural Hazards And Technological Accidents In Europe. An Overview Of The Last Decade*. EEA Technical report; 2010. p. 140.

[9]   Kang Y., Zhao C., Zhang Q., *et al*. 'Application of InSAR techniques to an analysis of the Guanling landslide'. *Remote Sensing*. 2017;**9**(10):1046.

[10] Tofani V., Raspini F., Catani F. 'Persistent scatterer interferometry (PSI) technique for landslide characterization and monitoring'. *Remote Sensing*. 2013;**5**(3):1045–65.

[11] Bardi F., Frodella W., Ciampalini A., *et al*. 'Integration between ground based and satellite SAR data in landslide mapping: the San Fratello case study'. *Geomorphology*. 2014;**223**(4):45–60.

[12] Ullo S.L., Addabbo P., Martire D.D., *et al*. 'Application of DInSAR technique to high coherence sentinel-1 images for dam monitoring and result validation through i*n situ* measurements'. *IEEE Journal of Selected Topics in Applied Earth Observations and Remote Sensing*. 2019;**12**(3):875–90.

[13] Addabbo P. 'Land cover classification and monitoring through multisensor image and data combination'. 2016 IEEE International Geoscience and Remote Sensing Symposium (IGARSS); Beijing, China; 2016. pp. 902–5.

[14] Fiscante N., Focareta M., Galdi C., *et al*. 'Analysis and validation of high-resolution satellite DEMs generated from EROS-B data for Montaguto landslide'. *2012 IEEE International Geoscience and Remote Sensing Symposium*. Munich, Germany: IGARSS; 2012. pp. 7549–52.

[15] Das I., Sahoo S., van Westen C., *et al.* 'Landslide susceptibility assessment using logistic regression and its comparison with a rock mass classification system, along a road section in the Northern Himalayas (India)'. *Geomorphology.* 2010;**114**(4):627–37.

[16] Akgun A. 'A comparison of landslide susceptibility maps produced by logistic regression, multi-criteria decision, and likelihood ratio methods: a case study at İzmir, Turkey'. *Landslides.* 2012;**9**(1):93–106.

[17] Pham B.T., Pradhan B., Tien Bui D., *et al.* 'A comparative study of different machine learning methods for landslide susceptibility assessment: a case study of Uttarakhand area (India)'. *Environmental Modelling & Software.* 2016;**84**(1):240–50.

[18] Kavzoglu T., Sahin E.K., Colkesen I. 'Landslide susceptibility mapping using GIS-based multi-criteria decision analysis, support vector machines, and logistic regression'. *Landslides.* 2014;**11**(3):425–39.

[19] Peng L., Niu R., Huang B., *et al.* 'Landslide susceptibility mapping based on rough set theory and support vector machines: a case of the three Gorges area, China'. *Geomorphology.* 2014;**204**:287–301.

[20] Cortes C., Vapnik V. 'Support-vector networks'. *Machine Learning.* 1995;**20**(3):273–97.

[21] Ullo S.L., Langenkamp M.S., Oikarinen T.P., *et al.* 'Landslide geohazard assessment with convolutional neural networks using sentinel-2 imagery data.'2019 IEEE International Geoscience and Remote Sensing Symposium; 2019. pp. 9646–9.

[22] Prakash N., Manconi A., Loew S. 'Mapping landslides on eo data: performance of deep learning models vs. traditional machine learning models'. *Remote Sensing.* 2020;**12**(12):346.

[23] Varnes D.J. *Slope movement types and processes [online].* 1978. Available from https://trid.trb.org/view/86168.

[24] Cruden D.M., Varnes D.J. 'U.S. National academy of sciences: landslide types and processes'. *Transportation Research Board, Special Report.* 1996;**247**:36–75.

[25] *Google Earth Engine*[online]. Available from https://earthengine.google.com/ [Accessed 10 Jan 2019].

[26] *NASA Global Landslide Catalog [online].* Available from https://data.nasa.gov/Earth-Science/Global-Landslide-Catalog/h9d8-neg4/data [Accessed 2 Jan 2019].

[27] Goodfellow I., Bengio Y., Courville A. *Deep Learning.* The MIT Press; 2016.

[28] Boureau Y.L., Ponce J., LeCun Y. 'A theoretical analysis of feature pooling in visual recognition'. Proceedings of the 27th International Conference on Machine Learning (ICML-10); 2010. pp. 111–18.

[29] Boureau Y.L., Le Roux N., Bach F. *Ask the locals: multi-way local pooling for image recognition.* International Conference on Computer Vision; Barcelona, Spain, 6-13 November 2011; 2011. pp. 2651–8.

[30] Malinowski M., Fritz M. 'Learnable pooling regions for image classification'. *arXiv preprint arXiv.* 2013;**13013516**.

[31]   Diederik P.K., Ba J. *Adam: a method for stochastic optimization [online].* *CoRR. 2014;abs/1412.6980.* Available from http://arxiv.org/abs/1412.6980.

[32]   Kingma D.P., Ba J. 'Adam: a method for stochastic optimization'. *arXiv.* 2014.

[33]   Mou L., Bruzzone L., Zhu X.X. 'Learning spectral-spatial-temporal features via a recurrent convolutional neural network for change detection in multi-spectral imagery'. *IEEE Transactions on Geoscience and Remote Sensing.* 2019;**57**(2):924–35.

[34]   Dozat T. *Incorporating nesterov momentum into adam.* 2019. Available from http://cs229.stanford.edu/proj2015/054_report.pdf.

[35]   Sutskever I., Martens J., Dahl G., *et al.* 'On the importance of initialization and momentum in deep learning'. *IEEE Int. Conf. Mach. Learn;* 2013. pp. 1–9.

[36]   Glorot X., Bengio Y. 'Understanding the difficulty of training deep feedforward neural networks'. *Journal of Machine Learning Research – Proceedings Track.* 2010;**01**(9):249–56.

[37]   NASA. *Sentinel-2 band [online].* 2020. Available from https://landsat.gsfc. nasa.gov/sentinel-2b-successfully-launches/.

[38]   Kohavi R. 'A study of cross-validation and bootstrap for accuracy estimation and model selection'. Vol. 14. Ijcai; Montreal, Canada; 1995. pp. 1137–45.

[39]   Mastery M.L. *K-fold cross validation [online].* 2020. Available from https:// machinelearningmastery.com/k-fold-cross-validation/.

[40]   Brodersen K.H., Ong C.S., Stephan K.E. 'The Balanced Accuracy and Its Posterior Distribution'. 2010 20th International Conference on Pattern Recognition; 2010. pp. 3121–4.

[41]   Hestness J., Narang S., Ardalani N., *et al. Deep learning scaling is predictable, empirically [online].* CoRR. 2017; abs/1712.00409. Available from http://arxiv.org/abs/1712.00409.

[42]   Mondini A., Santangelo M., Rocchetti M., *et al.* 'Sentinel-1 SAR amplitude imagery for rapid landslide detection'. *Remote Sensing.* 2019;**11**(7):760.

[43]   Zhao C., Liu X., Zhang Q. 'InSAR application to baige landslide event, china, from fast rescue to catchment investigation'. IGARSS 2019 - 2019 IEEE International Geoscience and Remote Sensing Symposium; 2019. pp. 9630–3.

[44]   Duan W., Wang C., Zhang H. 'Landslide detection and monitoring for moutainous areas of southwest China using time series InSAR'. IGARSS 2019 – 2019 IEEE International Geoscience and Remote Sensing Symposium; 2019. pp. 2038–41.

[45]   Sameen M.I., Pradhan B. 'Landslide detection using residual networks and the fusion of spectral and topographic information'. *IEEE Access.* 2019;**7**:114363–73.

[46]   *Sentinel user guides [online].* Available from https://sentinel.esa.int/web/sentinel/user-guides.

[47] Sebastianelli A., Rosso M.P.D., Ullo S.L. *Sentinel data downloader tool [online]. GitHub.* Available from https://github.com/Sebbyraft/SentinelData DownloaderTool.

[48] Google Colab [online]. 2021. Available from https://colab.research.google. com/notebooks/intro.ipynb#recent=true.

[49] *Tensorflow performance test: CPU VS GPU [online].* 2020. Available from https://medium.com/@andriylazorenko/ tensorflow-performance-test-cpu-vs-gpu-79fcd39170c.

[50] *Tensorflow. Keras accuracy [online].* 2020. Available from https://www.ten-sorflow.org/api_docs/python/tf/keras/metrics/Accuracy.

[51] Simone G., Farina A., Morabito F.C., *et al.* 'Image fusion techniques for remote sensing applications'. *Information Fusion.* 2002;**3**(1):3–15.

[52] Serpico S.B., Roli F. 'Classification of multisensor remote-sensing images by structured neural networks'. *IEEE Transactions on Geoscience and Remote Sensing.* 1995;**33**(3):562–78.

[53] Bruzzone L., Prieto D.F., Serpico S.B. 'A neural-statistical approach to multitemporal and multisource remote-sensing image classification'. *IEEE Transactions on Geoscience and Remote Sensing.* 1999;**37**(3):1350–9.

[54] Del Rosso M.P., Sebastianelli A., Ullo S.L. *Artificial Intelligence Applied to Satellite-based Remote Sensing Data for Earth Observation.* Book undercontract with a scheduled publication of spring/summer. IET (Institute ofEngineering and Technology) Digital Library; 2021.

[55] Sanchez E.H., Serrurier M., Ortner M. *Learning disentangled representations of satellite image time series [online].* CoRR. 2019;abs/1903.08863. Available from http://arxiv.org/abs/1903.08863.

[56] Hamaguchi R., Sakurada K., Nakamura R. *Rare event detection using disentangled representation learning.* CoRR. 2018;abs/1812.01285[online]. Available from http://arxiv.org/abs/1812.01285.

[57] ZY L., Shi W., Zhang X., *et al.* 'Landslide inventory mapping from bitemporal high-resolution remote sensing images using change detection and multiscale segmentation'. *IEEE Journal of Selected Topics in Applied Earth Observations and Remote Sensing.* 2018;**11**(5):1520–32.

[58] Ding A., Zhang Q., Zhou X. 'Automatic recognition of landslide based on CNN and texture change detection'. 2016 31st Youth Academic Annual Conference of Chinese Association of Automation (YAC); 2016. pp. 444–8.

[59] Vollrath A. *Open SAR toolkit [online].* GitHub. 2020. Available from https:// github.com/ESA-PhiLab/OpenSarToolkit.

[60] Correa-Muñoz N.A., Murillo-Feo C.A. 'Detection of landslides with SAR polarimetry: case study of south-eastern Colombia'. *ISPRS – International Archives of the Photogrammetry, Remote Sensing and Spatial Information Sciences.* 2018;**XLII-3/W4**:177–84.

[61] Orynbaikyzy A., Gessner U., Conrad C. 'Crop type classification using a combination of optical and radar remote sensing data: a review'. *International Journal of Remote Sensing.* 2019;**40**(17):6553–95.

[62] Kussul N., Lavreniuk M., Skakun S., *et al.* 'Deep learning classification of land cover and crop types using remote sensing data'. *IEEE Geoscience and Remote Sensing Letters.* 2017;**14**(5):778–82.

[63] Ji S., Zhang C., Xu A., *et al.* '3D convolutional neural networks for crop classification with multi-temporal remote sensing images'. *Remote Sensing.* 2018;**10**(1):75.

[64] Nowakowski A., Mrziglod J., Spiller D., *et al.* 'Crop type mapping by using transfer learning'. *International Journal of Applied Earth Observation and Geoinformation.* 2021;**98**:102313.

[65] Szegedy C., Liu W., Jia Y. 'Going deeper with convolutions'. Proceedings of the IEEE conference on computer vision and pattern recognition; 2015. pp. 1–9.

[66] Simonyan K., Zisserman A. 'Very deep convolutional networks for large-scale image recognition'. *arXiv preprint arXiv.* 2014;**14091556**.

[67] Pan S.J., Yang Q. 'A survey on transfer learning'. *IEEE Transactions on Knowledge and Data Engineering.* 2009;**22**(10):1345–59.

[68] Deng J., Dong W., Socher R. 'Imagenet: A large-scale hierarchical image database'. *2009 IEEE conference on computer vision and pattern recognition*; Ieee; 2009. pp. 248–55.

[69] Sumbul G., Kang J., Kreuziger T., *et al.* 'BigEarthNet dataset with a new class-nomenclature for remote sensing image understanding'. *arXiv preprint arXiv.* 2020;**200106372**.

[70] Blaschke T. 'Object based image analysis for remote sensing'. *ISPRS Journal of Photogrammetry and Remote Sensing.* 2010;**65**(1):2–16.

[71] Walter V. 'Object-based classification of remote sensing data for change detection'. *ISPRS Journal of Photogrammetry and Remote Sensing.* 2004;**58**(3):225–38.

[72] Zhou W., Troy A. 'An object-oriented approach for analyzing and characterizing urban landscape at the Parcel level'. *International Journal of Remote Sensing.* 2008;**06**(29):3119–35.

[73] Myint S.W., Gober P., Brazel A., *et al.* 'Per-pixel vs. object-based classification of urban land cover extraction using high spatial resolution imagery'. *Remote Sensing of Environment.* 2011;**115**(5):1145–61.

[74] Schiewe J. 'Segmentation of high-resolution remotely sensed data – concepts, applications and problems'. *International Archives of Photogrammetry and Remote Sensing.* 2012:34.

[75] DigitalGlobe Website [online]. 2020. Available from https://www.digital-globe.com/.

[76] Lourenço P., Teodoro A.C., Gonçalves J.A., *et al.* 'Assessing the performance of different OBIA software approaches for mapping invasive alien plants along roads with remote sensing data'. *International Journal of Applied Earth Observation and Geoinformation.* 2021;**95**(9):102263.

[77] Yan J., Zhou W., Han L., Qian Y. 'Mapping vegetation functional types in urban areas with WorldView-2 imagery: integrating object-based classification with phenology'. *Urban Forestry & Urban Greening.* 2018;**31**(3):230–40.

[78] Zhang C., Xie Z. 'Combining object-based texture measures with a neural network for vegetation mapping in the Everglades from hyperspectral imagery'. *Remote Sensing of Environment.* 2012;**124**:310–20.

[79] Lichtblau E., Oswald C.J. 'Classification of impervious land-use features using object-based image analysis and data fusion'. *Computers, Environment and Urban Systems.* 2019;**75**(2):103–16.

[80] Jahjah M., Ulivieri C. 'Automatic archaeological feature extraction from satellite VHR images'. *Acta Astronautica.* 2010;**66**(9):1302–10.

[81] El-naggar A.M. 'Determination of optimum segmentation parameter values for extracting building from remote sensing images'. *Alexandria Engineering Journal.* 2018;**57**(4):3089–97.

[82] Tomljenovic I., Tiede D., Blaschke T. 'A building extraction approach for airborne laser scanner data utilizing the object based image analysis paradigm'. *International Journal of Applied Earth Observation and Geoinformation.* 2016;**52**(2–3):137–48.

[83] Ullo S.L., Zarro C., Wojtowicz K., *et al.* 'LiDAR-based system and optical VHR data for building detection and mapping'. *Sensors.* 2020;**20**(5):1285. February;20:1285.

[84] eCognitin Website [online]. 2020. Available from https://docs.ecognition. com/v10.0.0/.

[85] Geoportale Nazionale Website [online]. 2020. Available from http://www. pcn.minambiente.it/mattm/progetto-piano-straordinario-di-telerilevamento/.

[86] Zhou W. 'An object-based approach for urban land cover classification: integrating LiDAR height and intensity data'. *IEEE Geoscience and Remote Sensing Letters.* 2013;**10**(4):928–31.

[87] Mesner N., Oštir K. 'Investigating the impact of spatial and spectral resolution of satellite images on segmentation quality'. *Journal of Applied Remote Sensing.* 2014;**8**(8):083696.

[88] Chirici G., Corona P., Travaglini D., *et al.* Segmentazione multirisoluzione e classificazione object oriented di immagini telerilevate. 2003.

[89] Awrangjeb M., Fraser C. 'Automatic segmentation of raw LIDAR data for extraction of building Roofs'. *Remote Sensing.* 2014;**6**(6):3716–51.

[90] Drăguţ L., Csillik O., Eisank C., *et al.* 'Automated parameterisation for multi-scale image segmentation on multiple layers'. *ISPRS Journal of Photogrammetry and Remote Sensing.* 2014;**88**(100):119–27.

[91] Su T. 'Scale-variable region-merging for high resolution remote sensing image segmentation'. *ISPRS Journal of Photogrammetry and Remote Sensing.* 2019;**147**:319–34.

[92] Zhao W., Du S., Emery W.J. 'Object-based convolutional neural network for high-resolution imagery classification'. *IEEE Journal of Selected Topics in Applied Earth Observations and Remote Sensing.* 2017;**10**(7):3386–96.

[93] Rajesh S., Nisia T., Selvaraj A. 'Land cover/land use mapping of LISS IV imagery using object-based Convolutional neural network with deep features'. *Journal of the Indian Society of Remote Sensing.* 2019;**48**(10).

[94]   Zhang C., Sargent I., Pan X., *et al.* 'An object-based convolutional neural network (OCNN) for urban land use classification'. *Remote Sensing of Environment*. 2018;**216**:57–70.

Chapter 9

A generation problem

Alessandro Sebastianelli[1], Maria Pia Del Rosso[1], Silvia Liberata Ullo[1], Erika Puglisi[2], and Filippo Biondi[3]

In this chapter we address the problem of generating satellite images, of a particular domain, using data of another domain. The description of the type of network used will be followed by a case study in which optical images are generated from synthetic aperture radar (SAR) data.

9.1 Introduction

The advent of high-resolution and high revision time satellites has led in recent years to the creation of huge datasets suitable for generative problems. The main idea is based on the use of a model, which may or may not be based on artificial intelligence, which is capable of generating new synthetic data. Among the main reasons, there is often the need to increase the size of a dataset, typically to increase the samples of an under-determined class. This phenomenon can happen, for example, when working with classification models. Sometimes it may happen that there are some classes with fewer examples than others; this involves an imbalance in the dataset and consequently an imbalance in the predictive power of the model. A possible solution to this problem can be the use of a generative model, which is able to synthetically generate the missing data and restore a balance in the dataset [1, 2].

In the remote sensing field, the generative networks can also be used to transfer information from one domain to another, with the aim of transferring the best characteristics from the input to the output. An interesting example is the generation of optical images using SAR data. Since SAR data are immune to weather and light conditions, it is possible to re-generate optical-like images corrupted by clouds or light conditions [3, 4]. In this chapter a generative adversarial network (GAN) for SAR to optical translation will be presented.

[1]Engineering Department, University of Sannio, Benevento, Italy
[2]Φ-lab, European Space Agency (ESA), Frascati, Italy
[3]Joint Satellite Remote Sensing Center, Italian Ministry of Defence, Rome, Italy

9.2   Deep generative models

With the spread of deep learning, a new family of methods, called deep generative models from the combination of generative models and deep neural networks, start to be used. It is possible to define a generative network like a learning model, part of unsupervised learning techniques in machine learning, with the aim of learning and emulating any kind of data distribution. All types of generative models aim at learning the true data distribution of the training set so as to generate new data with some variations. Obviously it is not always possible to learn the exact distribution of the dataset under analysis, so the idea is to model a distribution which is as similar as possible to the true data distribution. Thanks to the small number of parameters of these models with respect to the amount of data used during the training phase, the model is able to learn a function which can approximate the model distribution to the true distribution and learn the natural features of a dataset.

Two of the most common approaches are variational autoencoders (VAEs) and GANs. A VAE aims at maximizing the lower bound of the data log likelihood and a GAN aims at achieving an equilibrium between generator and discriminator.

9.2.1   *Variational autoencoders*

An autoencoder is a type of neural network composed of an encoder and a decoder, setted as neural networks, whose scope is to learn the best encoding–decoding scheme using an optimization process. The encoder compresses and maps the data into a latent[a] space Z in order to reduce the dimensions, and the decoder takes as input the representation Z and uses it to reconstruct the original input data (Figure 9.1).

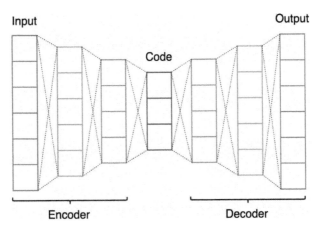

Figure 9.1   *Example of an autoencoder structure*

[a]Not directly observed but inferred through a mathematical model from other variables that are directly measured.

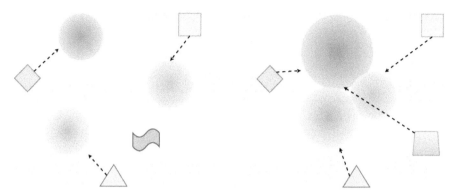

*Figure 9.2   Example of how the dataset is distributed in the space from a
classical autoencoder (on the left) and from a VAE (on the right).
In the first case the figure in purple taken as input will generate an
unrealistic output.*

Unlike classical autoencoders, VAEs are probabilistic generative models
defined in 2013 by Kingma *et al.* [5] and Rezende *et al.* [6]. The principal difference
between them is that the latent space in which the classical autoencoders convert
their inputs may not be continuous. This means that in generative models, where it
is necessary to generate variations from an input image, the decoder will generate
unrealistic outputs because during the training phase it never saw encoded vectors
coming from that particular region of space. VAEs, instead, have continuous latent
spaces, allowing easy random sampling and interpolation. The trick is in the encoder
architecture that is modeled to output two vectors instead of only one: a vector of
means (μ) and a vector of standard deviations (σ), which are used to obtain a sampled and varying encoding. Intuitively, the vector μ controls where the encoding
of an input should be centered around, while σ controls how much from the mean
the encoding can vary. In this way the decoder learns that not only a single point
in latent space refers to a sample of a particular class, but also all the other nearby
points (a distribution over the latent space) [7]. The final result will show encodings
as close as possible to each other, but still distinct, and this allows smooth interpolation and enables the construction of new samples (see Figure 9.2).

This result is achieved thanks to the introduction of the Kullback–Leibler (KL)
divergence into the loss function of VAEs. The KL divergence between two probability distributions measures how much they diverge from each other and, for VAEs,
is equivalent to the sum of all the KL divergences between the component X_i (with
mean μ_i and standard deviation σ_i^2) and the standard normal:

$$\text{KL} = \sum_{i=1}^{n} \sigma_i^2 + \mu_i^2 - \log(\sigma_i) - 1 \tag{9.1}$$

This formula is minimized when $\mu_i = 0$ and $\sigma_i = 1$. Its minimization forces the
encoder to distribute the encodings evenly around the center of the latent space and,
if it tries to cluster them into specific regions, it will be penalized. Regarding the

structures of the encoder and the decoder, they can be changed depending on the final goal of the network, e.g., to produce a synthetic human face photos, a convolutional–deconvolutional encoder–decoder pair can be used, and for sequential data, an LSTM (long short-term memory network) encoder–decoder pair can be used.

9.2.2   Generative adversarial networks

GANs are a model architecture for training a generative model described for the first time in 2014 by Ian Goodfellow [8] and then formalized by Alec Radford with a more stable approach called deep convolutional GANs [9]. The GAN architecture consists of two sub-models: a generator model for generating new examples and a discriminator model for classifying whether generated examples are real (from the domain) or fake (generated by the generator model). The two models are trained together in an adversarial zero-sum game[b] until the error rate of the discriminator model starts to increase and it turns out to be fooled about half the time (this means that the generator model is generating plausible examples that the discriminator is no more able to recognize as fake). The generator model takes as inputs N-dimensional vectors which represent the type of data it is of interest to generate. Since similar types of data will give dimensional vectors similarly distributed according to a probability distribution over the entire N-dimensional vector space, it is possible to say that the problem of generating a particular type of data is equivalent to the problem of generating new vectors with respect to a specific probability distribution. So, the final idea of a generator model is to take as input an N-dimensional uniform random variable and return as output another N-dimensional random variable that should follow the same probability distribution [10]. To train this kind of network, two methods should be followed: the first one consists in comparing the true and the generated probability distributions and backpropagating the error through the network (this method is used by generative matching networks [11]); the second one, used by GANs, is to train the model enforcing the generated distribution to be close to the true distribution, so the discriminator model will not be able to distinguish them. The discriminator model takes as input an N-dimensional variable and returns as output the probability of this variable to be a real one. In Figure 9.3 the GAN architecture is represented by a schema.

The architectures of the generator and the discriminator model can change with respect to the final goal of the GAN. In fact, different types of GAN exist, each one with a different scope and with a different architecture, e.g., a CycleGAN can transform an image from one domain to another [12], a text-to-image GAN can create an image from a sentence [13], and so on. Anyway, all the GANs have common problems which are areas of active researches:

---

[b]A zero-sum game is a mathematical representation of a situation in which each participant's gain/loss of utility is exactly balanced by the losses/gains of the utility of the other participants. In fact, in this case, the generator model's goal is to maximize the final classification error, while the discriminator model's goal is to minimize the final classification error.

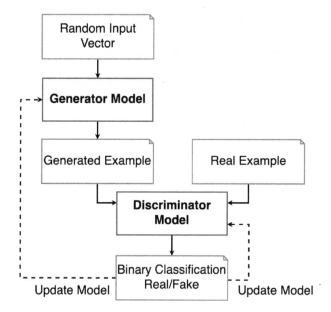

Figure 9.3   Example of a GAN architecture

- Vanishing gradients (see Section 4.4.7): When the discriminator model is too good, then the generator training can fail due to vanishing gradients because an optimal discriminator does not provide enough information to make progress. An attempt to remedy can be trying to use the Wasserstein loss.
- Mode collapse: It happens when a generator model is only capable of generating one or a small subset of different outcomes. An attempt to remedy can be trying to use an unrolled GAN [14].
- GANs often do not converge: Some forms of regularization can be used to try to improve GAN convergence, e.g., adding noise to discriminator inputs [15] or penalizing discriminator weights [16].
- Unbalance between the generator and discriminator causes overfitting.
- GANs are highly sensitive to the hyperparameter selections.

In [17], a lot of evaluating methods applied to GANs are presented with their pros and cons.

9.3   Case study: Generation of optical-like images

In this section, the implementation of a deep learning model for the generation of optical data will be presented. Two approaches will be introduced: in the first one, Sentinel-1 data are emulated using Sentinel-2 data; in the second one, only Sentinel-1 data are used [18].

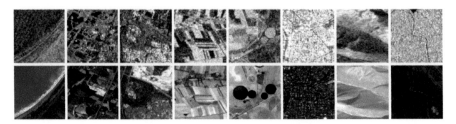

Figure 9.4   Samples from the dataset

9.3.1   Dataset and proposed methods

In this case study, the SEN1-2 dataset will be used. This dataset contains 282 384 pairs of SAR and optical image patches acquired by the Sentinel-1 and Sentinel-2 satellites, respectively. The patches are randomly distributed across the land masses of the Earth and spread over all four meteorological seasons. This is reflected by the dataset structure. For the SAR patches, 8-bit single-channel images representing the sigma0 backscatter value in dB scale are provided. For the optical patches, 8-bit color images representing the bands 4, 3, and 2 are used.

The SEN1-2 dataset is shared under the open-access license CCBY and available for download at a persistent link provided by the library of the Technical University of Munich: https://mediatum.ub.tum.de/1436631 [19]. Figure 9.4 shows a series of example images taken from the dataset.

In general the idea is to use speckle-free Sentinel-1 images to generate Sentinel-2 images, as shown in Figure 9.5.

The first approach used for the generation of optical data involves the use of simulated SAR data. In this case, the Sentinel-1 data are substituted with Sentinel-2 gray scale images. In this configuration, the network will only learn how to colorize the gray scale image. However, since the dataset contains Sentinel-2 and Sentinel-1 images, the second and more natural approach involves the use of real SAR data for the generation of optical ones.

SENTINEL 1   SENTINEL 1 FILTERED   SENTINEL 2 GENERATED

Figure 9.5   Sentinel-2 data generation

9.4 Model for the generation of optical-like data

The proposed model consists of two sub-models: the generator and the discriminator. In general, the proposed model belongs to a family of neural networks called GANs [8]. The generator takes as input a filtered SAR image and tries to generate an optical-like image. In this case the generator is an encoder–decoder structure called U-net, often used for segmentation tasks [20–22]. The discriminator is a classifier that takes as input two images, the real optical image and the generated one, and tries to classify them. The equilibrium is reached when the discriminator cannot distinguish the generated from the real image anymore.

9.4.1 Generator

As stated before, the proposed architecture for the generator is based on the U-net. This structure can be divided into two branches. The one on the left is a series of convolutional layers with stride 1×1, kernel size 4×4, and the number of filters gradually descending, from 64 to 16, followed by a rectified linear unit (ReLU) activation function and a batch normalization layer. The one on the right is a series of convolutional layers of the same type as the previous case but preceded by an up-convolution layer, which ensures that the dimensionality of the data is concordant for each level. In fact, for each level between the left branch and the right branch, there is a skip connection. The network ends in a three-dimensional convolutional layer so that its output is a color image (RGB). The generator summary scheme is shown in Figure 9.6.

This type of network, through a reduction of the features map, is able to learn small details of the image. The skip connections instead help the network to learn general details of the image, so in the up-convolution phase the network is able to associate a color image with a gray scale image. It is also able to simulate the spatial geometry of a Sentinel-2 image from a Sentinel-1 image.

9.4.2 Discriminator

The proposed architecture for the discriminator is based on a particular type of classifier. In particular, the structure consists of five convolutional layers, with a growing number of filters from 64 to 512 up to the penultimate layer, and with the last layer having only one filter. For all layers the stride is 1×1 and the kernel size is 4×4. The first and the last layers do not have batch normalization, but all except the last one are followed by the LeakyReLU activation function. The network input is the concatenation of the real Sentinel-2 image and the image generated by the generator, instead of the output, is a prediction; the network tends to predict whether the generated image is almost real or not. The discriminator summary scheme is shown in Figure 9.7.

In this particular network the predictive layer, which is usually dense, is replaced by a convolutional layer.

9.4.3 Loss function selection

The loss function is crucial, as it guides the gradient descent toward the minimum. Compared to the previous chapter architectures, the networks in this case are two,

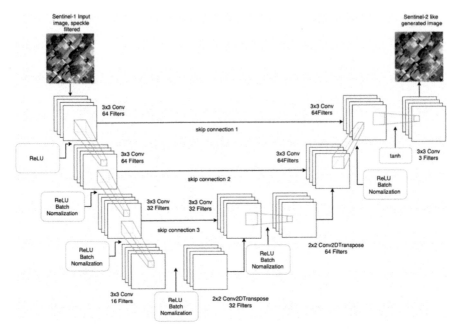

Figure 9.6   Network architecture for the generator

so two loss functions are required. Previous work on this topic showed that for the generator a good loss function is the mean absolute error and for the discriminator the mean squared error [23].

9.5   Results

The model was trained for seven epochs (for each epoch the model was trained on about 300 batches) using the Adam optimizer, with an initial learning rate of 0.0002.

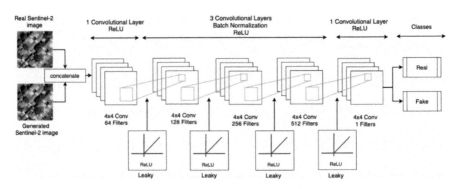

Figure 9.7   Network architecture for the discriminator

The discriminator model has a number of parameters to be optimized equals to 5 536 898 of which 2 769 345 are not trainable. The generator model has a number of parameters to be optimized equals to 156 915 of which only 544 are not trainable.

The training dataset (a small part of the original one) is composed of about 10 000 Sentinel-2 images and 10 000 Sentinel-1 images with speckle. The dataset is subjected to data augmentation to artificially increase, through transformations of the starting images, the actual dimension of the dataset. During each epoch a batch of 24 images, 12 for Sentinel-1 and 12 for Sentinel-2, is randomly taken from the memory (both for training and validation) through a "image generator." The generator loads the images, normalizes them, applies for each Sentinel-1 image the speckle filter, applies transformations (rotation, shift, etc.) and yields (returns) both the Sentinel-2 images (y train) and the "noise-free" Sentinel-1 images (x train).

9.5.1   Test on simulated Sentinel-1 images

As a first result, the network was trained and tested on simulated Sentinel-1 images; in practice, to totally exclude the speckle, the Sentinel-2 images were first transformed into gray scale. In this way the gray scale images represent the Sentinel-1 data and the RGB the Sentinel-2 data. The only effort of the network is the association of the correct colors, without having to reconstruct the geometry of the scene. The network used is the one described in the previous section, except that in this case the image generator must perform different steps. This network also requires minor computational efforts and less training time.

Figures 9.8–9.10 show some results obtained with this network. The figures should be analyzed per rows, since each row shows respectively the model input (the Sentinel-1 image), the ground truth (the Sentinel-2 images), and the model output (the generated Sentinel-2 image).

9.5.2   Test on real Sentinel-1 images

In this case the model was trained and tested with real Sentinel-1 images. In this case the network has to perform more complex operations, since there exists evident difference between SAR and optical images. Figures 9.11–9.13 show some results obtained with this network. The figures should be analyzed per rows, since each row shows respectively the model input (the Sentinel-1 image), the ground truth (the Sentinel-2 images), and the model output (the generated Sentinel-2 image).

Due to the complexity of the task, the results are quite different with respect to the ground truth, especially for the colors assignment. To improve the results, the authors are working on a new solution based on the use of time series for both Sentinel-1 and Sentinel-2 images.

9.6   Conclusions

The project proposed in this chapter is currently evolving. The latest idea is to add the dimension of time to data. With the introduction of the time series, the network

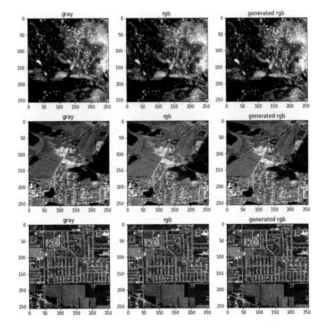

Figure 9.8   Test of the model on simulated Sentinel-1 images (1)

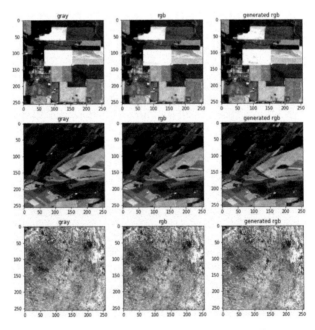

Figure 9.9   Test of the model on simulated Sentinel-1 images (2)

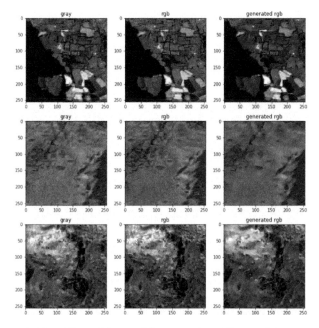

Figure 9.10   Test of the model on simulated Sentinel-1 images (3)

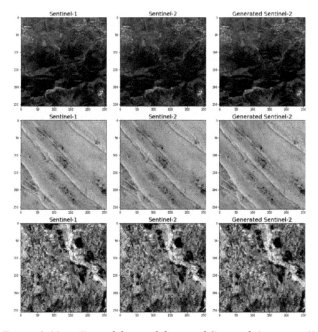

Figure 9.11   Test of the model on real Sentinel-1 images (1)

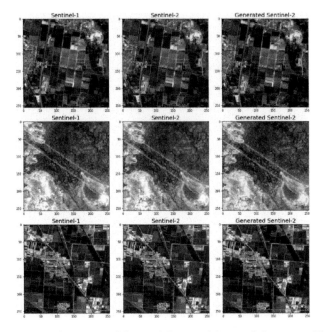

Figure 9.12   Test of the model on real Sentinel-1 images (2)

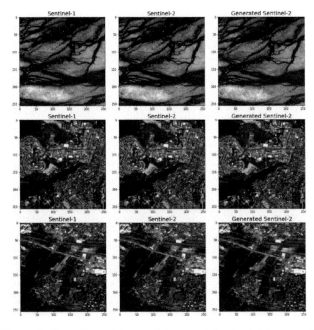

Figure 9.13   Test of the model on real Sentinel-1 images (3)

should be able to "reconstruct" the optical-like image, constrained to temporal features. For instance, the network can learn the seasonal evolution of the scene and then can produce an image with specific temporal-related features. The dataset has been built using the tool presented in [24].

References

[1]   Ali-Gombe A., Elyan E., Jayne C. 'Multiple fake classes GAN for data augmentation in face image dataset'. *2019 International Joint Conference on Neural Networks (IJCNN)*. IEEE; 2019. pp. 1–8.

[2]   Ali-Gombe A., Elyan E. 'MFC-GAN: class-imbalanced dataset classification using multiple fake class generative adversarial network'. *Neurocomputing*. 2019;**361**(4):212–21.

[3]   Schmitt M., Hughes L.H., Zhu X.X. 'The SEN1-2 dataset for deep learning in SAR-optical data fusion'. *arXiv*. 2018.

[4]   Wang L., Xu X., Yu Y., *et al.* 'SAR-to-optical image translation using supervised cycle-consistent adversarial networks'. *IEEE Access*. 2019;7:129136–49.

[5]   Kingma D.P., Welling M. 'Auto-encoding variational bayes'. *arXiv*. 2013.

[6]   Rezende D.J., Mohamed S., Wierstra D. 'Stochastic backpropagation and approximate inference in deep generative models'. *arXiv*. 2014.

[7]   Shafkat I. *Intuitively understanding variational autoencoders [online]*. 2018. Available from https://towardsdatascience com/intuitivelyunderstanding-variational-autoencoders-1bfe67eb5daf [Accessed November 2020].

[8]   Goodfellow I.J., Pouget-Abadie J., Mirza M., *et al.* 'Generative adversarial nets'. *Advances in Neural Information Processing Systems*; 2014.

[9]   Radford A., Metz L., Chintala S. 'Unsupervised representation learning with deep convolutional generative adversarial networks'. *arXiv*. 2015.

[10] Rocca J. Understanding generative adversarial networks (GANs). 2019. Available from https://towardsdatascience.com/understanding-generative-adversarial-networks-gans-cd6e4651a29.

[11] Bartunov S., Vetrov D.P. 'Fast adaptation in generative models with generative matching networks'. *arXiv preprint arXiv*. 2016;**161202192**.

[12] Zhu J.Y., Park T., Isola P., *et al.* 'Unpaired image-to-image translation using cycle-consistent adversarial networks'. *Proceedings of the IEEE international conference on computer vision*; 2017. pp. 2223–32.

[13] Reed S., Akata Z., Yan X., *et al.* 'Generative adversarial text to image synthesis'. *arXiv*. 2016.

[14] Metz L., Poole B., Pfau D., *et al.* 'Unrolled generative adversarial networks'. *arXiv*. 2016.

[15] Arjovsky M., Bottou L. 'Towards principled methods for training generative adversarial networks'. *arXiv preprint arXiv*. 2017;**170104862**.

[16] Roth K., Lucchi A., Nowozin S., *et al.* 'Stabilizing Training of Generative Adversarial Networks Through Regularization'. *Advances in Neural Information Processing Systems*; 2017. pp. 2018–28.

[17]   Borji A. 'Pros and cons of gan evaluation measures'. *Computer Vision and Image Understanding*. 2019;**179**(1):41–65.

[18]   Sebastianelli A. 'SAR and Optical data fusion using Artificial Intelligence. Generation of Sentinel-2 images from despeckled Sentinel-1 data. Thesis discussion date: 17-10-2019'.

[19]   Schmitt M. *SEN1-2 [Dataset] [online]*. Technical University of Munich. 2018. Available from https://mediatum.ub.tum.de/1436631 [Accessed November 2020].

[20]   Ronneberger O., Fischer P., Brox T. 'Unet. MICCAI2015'. 2015.

[21]   Zhou Z., Rahman Siddiquee M.M., Tajbakhsh N., *et al.* 'Unet++: A nested U-Net architecture for medical image segmentation'. *Lecture Notes in Computer Science (including subseries Lecture Notes in Artificial Intelligence and Lecture Notes in Bioinformatics)*; 2018.

[22]   Zhou Z., Siddiquee M.M.R., Tajbakhsh N., Liang J. 'UNet++: redesigning SKIP connections to exploit multiscale features in image segmentation'. *IEEE Transactions on Medical Imaging*. 2020;**39**(6):1856–67.

[23]   Wang P., Patel V.M. 'Generating high quality visible images from SAR images using CNNs'. 2018 IEEE Radar Conference; 2018.

[24]   Sebastianelli A., Rosso M.P.D., Ullo S., *et al. Automatic dataset builder for machine learning applications to satellite imagery*. Available from https://arxiv.org/abs/2008.01578 [Accessed September 2020].

Chapter 10

A filtering problem: SAR speckle filtering

Alessandro Sebastianelli[1], Maria Pia Del Rosso[1], Silvia Liberata Ullo[1], Andrea Radius[2], Carmine Clemente[3], Danilo Orlando[4], Filippo Biondi[5], and James Wheeler[6]

This chapter presents the problem of speckle filtering. After an introduction to the problem, the authors propose a case study on speckle filtering from Sentinel-1 images, using a Deep Learning approach.

10.1 Introduction

Speckle is a granular interference that inherently exists in and degrades the quality of the active radar, synthetic aperture radar (SAR), medical ultrasound and optical coherence tomography images. The vast majority of surfaces, synthetic or natural, are extremely rough on the scale of the wavelength. Images obtained from these surfaces by coherent imaging systems such as laser, SAR, and ultrasound suffer from a common interference phenomenon called speckle. The origin of this phenomenon is seen if we model our reflectivity function as an array of scatterers. Because of the finite resolution, at any time the sensor is receiving reflections from a distribution of scatterers within the resolution cell. As these scattered signals add coherently, constructive and destructive interference phenomena are generated. Speckle is the result of these phenomena appearing as intensity fluctuation in the final image.

Although commonly referred to as 'speckle noise', speckle is not noise in its generally understood sense of an unwanted modification to a desired signal. Rather, it is the signal itself that fluctuates, because the scatterers are not identical for each cell, and the signal is highly sensitive to small variations in scatterers. Speckle in SAR is generally serious, causing difficulties for image interpretation. It is caused

[1]Engineering Department, University of Sannio, Benevento, Italy
[2]ICEYE, Espoo, Finland
[3]University Of Strathclyde, Glasgow, Scotland
[4]University Niccolò Cusano, Rome, Italy
[5]Joint Satellite Remote Sensing Center, Italian Ministry of Defence, Rome, Italy
[6]Φ-Lab, European Space Agency (ESA), Frascati, Italy

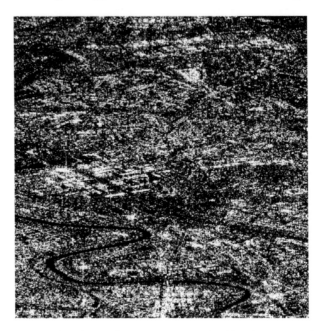

Figure 10.1   Example of a synthetic aperture radar (SAR) image with speckle (Rome, VV polarisation)

by coherent processing of backscattered signals from multiple distributed targets. In SAR oceanography, for example, speckle is caused by signals from elementary scatterers, the gravity-capillary ripples, and manifests as a pedestal image, beneath the image of the sea waves.

The noise of single-look SAR amplitude imagery is often modelled using the Rayleigh distribution, whereas that of single-look intensity imagery is modelled with a negative exponential distribution. Multilook SAR imagery usually follows a gamma distribution, assuming that the looks are independent [1, 2].

10.1.1   Speckle filtering motivation

Unlike optical remote sensing images, characterised by very clear and uniform features, SAR images are affected by speckle due to several scattering centres in each resolution cell whose returns interfere positively or negatively, generating light or dark image brightness. For this reason, speckle confers to SAR images a granular aspect with random spatial variations. An example of single polarisation speckled SAR image is shown in Figure 10.1.

Figure 10.1 shows the intensity values of the SAR image. This has a poor contrast and random aspect, due to the speckle. This phenomenon reduces the possibilities of visual analysis and interpretation of the scene under consideration. The classification of different media, by comparing the intensity value to a threshold, leads, typically, to various errors due to the high variability also introduced by speckle.

The speckle phenomenon also affects the phase of scattering coefficients and corrupts polarimetric[1] information [3].

10.1.2   *Speckle reduction*

Speckle reduction techniques can be classified into two categories: single product speckle filtering and multi-temporal speckle filtering. Several different methods for single product speckle filtering are used to reduce speckle, based upon different mathematical models of the phenomenon. The classic approach for speckle reduction is the multilook process. It implies the concept of look, or sub-aperture, that corresponds to the image generated using only a portion of the full bandwidth. The looks can be generated in range, taking a portion of the transmitted chirp bandwidth, and in azimuth, using part of the synthetic aperture. Every look can be considered independent from each other, and the looks can be summed incoherently to generate a multilook image. Multilook processing normally includes data sub-sampling and ground-projection, increasing radiometric resolution while decreasing spatial resolution. After multilooking processing, the phase information is lost. Multilooking could be also performed on amplitude/intensity images averaging the pixels in range and azimuth.

This approach, for example, employs multiple-look[2] processing (also known as multilook processing), averaging out the speckle by taking several 'looks' at a target in a single radar sweep. The average is the incoherent average of the looks.

Another approach involves using adaptive and non-adaptive filters for signal processing (where adaptive filters adapt their weightings across the image to the speckle level, and non-adaptive filters apply the same weightings uniformly across the entire image). This type of filtering degrades the image and eliminates information. The applicability of filtering and the choice of filter type involves compromises. Adaptive speckle filtering is better at preserving edges and detail in high-texture areas (such as forests or urban areas). Non-adaptive filtering is simpler to implement, and requires less computational power.

There are two typical forms of non-adaptive speckle filtering: one based on the mean and one based upon the median. The former is better at preserving edges while the latter eliminate the spikes than the former. There are many forms of adaptive

---

[1]Polarimetry is the measurement and interpretation of the polarisation of transverse waves, most notably electromagnetic waves, such as radio or light waves. Typically polarimetry is done on electromagnetic waves that have travelled through or have been reflected, refracted or diffracted by some material in order to characterise that object.

[2]The process of multilooking improves the SAR image quality by reducing the speckle and allows to obtain a square pixel on the output image. Either within range or azimuth direction or in both the directions, subsequent lines are averaged to get a better image. Averaging the different looks of one single image will result in a loss of resolution. For example, if the resolutions in azimuth and range directions of an SAR are 5 and 20, respectively, the image can be multilooked four times in azimuth direction so that the size of each pixel becomes 20 × 20.

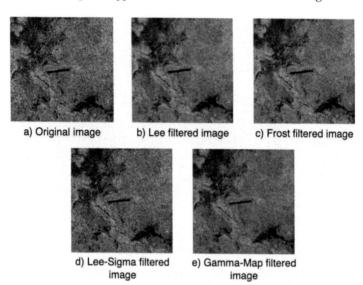

a) Original image   b) Lee filtered image   c) Frost filtered image

d) Lee-Sigma filtered   e) Gamma-Map filtered
image   image

Figure 10.2   Comparison of speckle filters (source [6])

speckle filtering, including the Lee filter, the Frost filter and the Refined Gamma Maximum-A-Posteriori (RGMAP) filter. They all rely upon three fundamental assumptions in their mathematical models, these assumptions are:

- Speckle in SAR is a multiplicative effect, i.e. it is in direct proportion to the local grey level in any area.
- The signal and the speckle are statistically independent of each other.
- The sample mean and variance of a single pixel are equal to the mean and variance of the local area that is centred on that pixel.

The Lee filter converts the multiplicative model into an additive one. In this case the problem is reduced to a known and easily tractable case [4]. This is a fundamental aspect of Lee filters and, as demonstrated in the following sections, it enables Deep Learning to be effective when dealing with speckle filtering.

The second category of speckle filtering techniques is the multitemporal filtering approach. Multitemporal filtering is applied on a dataset of co-registered images acquired with similar geometry. The filtering is performed averaging the independent images and allows to achieve a very high signal-to-noise ratio (SNR) without losing spatial resolution [5].

10.1.3   Speckle filtering examples

In this section classic speckle filters are presented. Figure 10.2 shows SAR images filtered using the most common filters discussed in the previous section.

In order to compare the qualitative results, expressed by the images in Figure 10.2, it is worth evaluating the results using some common metrics. The

Table 10.1   *Comparison of ENL, Robert's Edge Operator and Variance Texture Operator of the original image and the images filtered using the Lee, Frost, Lee-Sigma, Gamma-Map. A 5×5 window was used in all cases (source [6])*

Image	ENL	Robert's edge operator	Variance texture operator
Original image	3.64	82.85	1 512.83
Lee	11.07	19.28	145.07
Frost	9.40	23.79	244.36
Lee-Sigma	6.04	50.96	653.51
Gamma-Map	12.54	14.83	91.15

ENL[3], the Robert's Edge Operator[4] and the Variance Texture Operator[5] values of the original image and the filtered images are reported in Table 10.1.

10.2   Case study: Sentinel-1 speckle filtering

In this section the implementation of a Deep Learning model for speckle filtering will be presented [9]. Two approaches will be introduced; in the first approach Sentinel-1 data are emulated using Sentinel-2 data. While in the second approach only Sentinel-1 data are used. Both strategies involve the use of data without speckle as the ground truth and noisy artificially generated images as the input for the convolutional neural network (CNN) [10].

10.2.1   Multi-modal speckle filtering

In order to train a machine learning model, both the input and desired output are necessary when training it in a supervised manner. Since in this case the desired output is the Sentinel-1 speckle-free image, the first strategy proposes the use of Sentinel-2 images instead of Sentinel-1 data. By converting to grey scale a Sentinel-2 RGB (Red-Green-Blue) image, the ground-truth/desired output is created. Then the speckle is generated using the statistical model previously presented and by adding

---

[3]Equivalent number of looks (ENL) is often used to estimate the speckle noise level in a SAR image [4]. ENL is the mean-to-standard deviation ratio, which is a measure of the SNR and is often calculated in a homogeneous image area [4]. ENL is used to measure the degree of speckle reduction in this study. The higher the ENL value, the stronger the speckle reduction.

[4]The ENL can only be used to measure the speckle-smoothing capability of filters, because it carries no information on the loss of high-frequency information [4]. Thus, Robert's edge detector [7] was employed to measure the edge-preserving performances of the commonly used filters and the new filter.

[5]Texture information in remotely sensed imagery, especially SAR images, has received increasing attention in image interpretation and classification. The texture preserving capability, therefore, should play an important role in measuring the performance of a speckle filter. A second-order texture operator, Variance [8], was used to measure the retention of texture information in the original and the filtered images.

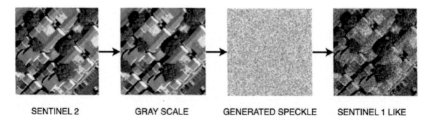

SENTINEL 2   GRAY SCALE   GENERATED SPECKLE   SENTINEL 1 LIKE

Figure 10.3   Generation of Sentinel-1 like images

the speckle to the ground-truth the input image is created. A Sentinel-1 like dataset can then be generated by repeating the abovementioned approach on the entire set of Sentinel-2 images. The procedure is sketched in Figure 10.3.

Even if this method was not found to be very adequate to real Sentinel-1 images, due in part to the obvious geometrical differences between Sentinel-1 and Sentinel 2 images and how the pixel values are distributed for the two types of data, it is worth highlighting that Sentinel-2 images are also affected by noise and so the procedure for creating the dataset can be very useful. The advantage of this procedure is that the user is able to create a dataset with pairs of images (with noise and without noise) starting from a smaller dataset (without noise) and generating from each image its twin (with noise).

10.2.2   Pure SAR strategy

The second method follows the same principle of the first one, but in this case only Sentinel-1 data is used. In this particular case, to create the ground-truth (the Sentinel-1 image without speckle) an average of a time series over a geographical area was used. In fact, as mentioned above, by taking the mean of the data acquired over the same area leads to an image without speckle. This choice was driven by the fact that the CNN should not be comparable to classic filters; in fact, instead of using the temporal average, the Lee filter or the multilook could be applied, but in this case the CNN will not lead to better results. To generate the input, the Sentinel-1 with speckle, the speckle was simply generated and applied to the ground-truth, as sketched in Figure 10.4. The dataset was implemented by repeating this procedure on different geographical areas.

This strategy has proven to be much more effective on real Sentinel-1 images [11]. The results obtained with this method will be presented in the section 10.4.

10.2.3   CNN model for SAR speckle filtering

The CNN model is composed of several convolutional layers with a constant number of filters that take care of extracting speckle noise from the input image. As for the Lee filter, the proposed model will convert the multiplicative model into an additive one. Indeed the CNN is used to estimate the noise. The filtered image is obtained by subtracting the estimated noise from the input image. The structure of the model

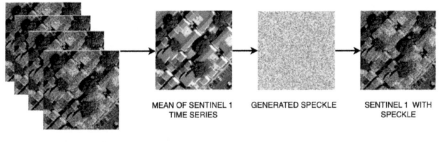

MEAN OF SENTINEL 1
TIME SERIES

GENERATED SPECKLE

SENTINEL 1 WITH
SPECKLE

SENTINEL 1 TIME SERIES

Figure 10.4   Generation of Sentinel-1 images with speckle

is flexible; in fact the number of convolutional layers can be varied based on the power with which the speckle noise is generated.

10.2.4   Network architecture

The architecture was designed to filter out speckle-type noise that affects Sentinel-1 images. In particular, it is composed of a first convolutional layer with a number of filters equal to 64, a 1×1 stride, a 3×3 kernel size, followed by the ReLU (Rectified Linear Unit) activation function. This first interface layer, that has as input the Sentinel-1 image with artificial speckle, is followed, in this particular case, by 13 macro layers. Each layer consists of a convolutional layer, with 64 filters each (the stride is 1×1 and the kernel size is 3×3), a batch normalisation layer and a ReLU layer. The central body of the network is the noise estimator, through the series of 13 layers that transform the two-dimensional input image into multidimensional data. At the end of the central block the data is again converted to the initial state (two-dimensional image and the speckle representative estimated by the CNN). The next layer subtracts the estimated noise from the input image and returns the filtered image. The network summary scheme is shown in Figure 10.5.

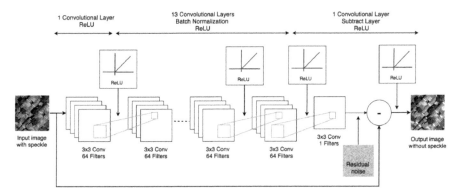

Figure 10.5   Network architecture for synthetic aperture radar (SAR) image despeckling

10.2.5   Loss function selection

The choice of a loss function is a crucial step for designing a neural network, especially in CNN-based image reconstruction tasks. Several works have explored different loss functions and their combinations for effective learning of tasks such as super-resolution, semantic segmentation and style transfer. Previous works on CNN-based image restoration optimised over pixel-wise L2-norm (Euclidean loss) or L1-norm between the predicted and ground truth images. In general the overall loss function should at least ensure that the recovered image contains pixel-level details and at the same time maintain the smoothness [12, 13].

For the task proposed in this chapter, a simple loss function is selected: the mean squared error (MSE). As the reader will see the MSE proved to be a sufficient starting point for the creation of a model-able of filter speckle noise.

10.3   Evaluation metrics

In order to verify the effective functioning of the network and to be able to define the quality of the filtered data, it is necessary to introduce some evaluation metrics in the discussion. In order to verify the effective functioning of the network and to be able to define the quality of the filtered data, it is necessary to introduce some evaluation metrics in the discussion. Subsequently, a metric was chosen to determine whether the filtering neural network is capable of preserving the structure of the image itself. The metrics listed above are the peak signal-to-noise ratio (PSNR) and the structural similarity index (SSIM).

10.3.1   Peak signal-to-noise ratio

PSNR is an engineering term for the ratio between the maximum possible power of a signal and the power of corrupting noise that affects the fidelity of its representation. Because many signals have a very wide dynamic range, PSNR is usually expressed in terms of the logarithmic decibel scale.

PSNR is easily defined using the MSE. Given a noise-free *mxn* monochrome image I and its noisy approximation K, MSE is defined as:

$$\text{MSE} = \frac{1}{mn} \sum_{i=0}^{m-1} \sum_{j=0}^{n-1} [I(i,j) - K(i,j)]^2 \qquad (10.1)$$

PSNR (in dB) is defined as:

$$
\begin{aligned}
\text{PSNR} \quad &= 10 \cdot \log_{10} \left(\frac{\text{MAX}_I^2}{\text{MSE}} \right) \\
&= 20 \cdot \log_{10} \left(\frac{\text{MAX}_1}{\sqrt{\text{MSE}}} \right) \\
&= 20 \cdot \log_{10}(\text{MAX}_1) - 10 \cdot \log_1 0 \, (\text{MSE})
\end{aligned}
\qquad (10.2)
$$

Here, MAX_I is the maximum possible pixel value of the image. When the pixels are represented using 8 bits per sample, this is 255. More generally, when samples are represented using linear PCM with B bits per sample, MAX_I is $2^B - 1$.

10.3.2 The structural similarity index

The SSIM is a method for predicting the perceived quality of digital television and cinematic pictures, as well as other kinds of digital images and videos.

SSIM is used for measuring the similarity between two images. The SSIM is a full reference metric; in other words, the measurement or prediction of image quality is based on an initial uncompressed or distortion-free image as reference.

SSIM is a perception-based model that considers image degradation as perceived change in structural information, while also incorporating important perceptual phenomena, including both luminance masking and contrast masking terms. Structural information is the idea that the pixels have strong interdependencies especially when they are spatially close. These dependencies carry important information about the structure of the objects in the visual scene. Luminance masking is a phenomenon whereby image distortions (in this context) tend to be less visible in bright regions, while contrast masking is a phenomenon whereby distortions become less visible where there is significant activity or 'texture' in the image.

The SSIM is calculated on various windows of an image. The measure between two windows x and y of common size NxN is:

$$\text{SSIM}(x,y) = \frac{(2\mu_x\mu_y + c_1)(2\sigma_{xy} + c_2)}{(\mu_x^2 + \mu_y^2 + c_1)(\sigma_x^2 + \sigma_y^2 + c_2)} \tag{10.3}$$

with μ_x the average of x, μ_y the average of y, σ_x^2 the variance of x, σ_y^2 the variance of y, σ_{xy} the covariance of x and y, $c_1 = (k_1 L)^2$ and $c_2 = (k_2 L)^2$ two variables to stabilise the division with weak denominator, L the dynamic range of the pixel values (typically this is $2^{\#bits\ per\ pixel} - 1$) and $k_1 = 0.01$ and $k_2 = 0.03$ by default.

10.4 Results

The model was trained for 60 epochs, using the Adam as the optimiser. The model to be trained has a number of parameters to be optimised equal to 484 609 of which only 1 664 are not trainable.

The training dataset is composed of about 3 000 images without speckle and 3 000 with speckle. The validation dataset instead is composed of about 600 images without speckle and 600 with speckle. Both datasets are subjected to data augmentation to artificially increase, through transformations of the starting image, the actual dimensions of the dataset.

Two types of dataset were implemented, one containing the averaged amplitudes and the other containing the averaged intensities [14]. During each epoch a batch of 12 Sentinel-1 images without speckle is randomly taken from the memory (both for training and validation) through a function that in slang is called 'generator'. The use of the generator allows to better manage the memory, in fact the data are loaded in dynamic memory in small batches and after being used for training these are replaced by a new batch. This technique is commonly used when working with GPUs (Graphics Processing Unit) or machines with little memory.

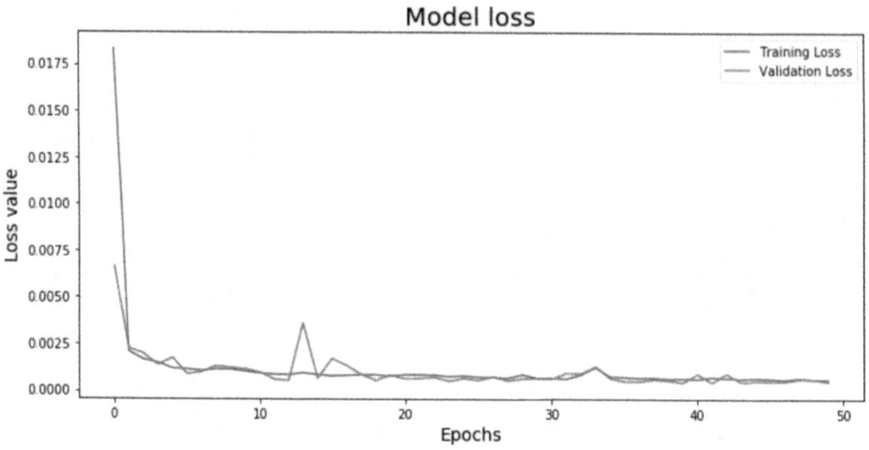

Figure 10.6　Model loss

The generator then loads the images, standardises and normalises them, generates for each one a speckle noise sample (Gamma distribution sample), produces a new image by multiplying the noise for the input image, applies transformations (rotation, shift, etc.) and yields (returns) both the starting image (y train) and the noisy image (x train).

Through this procedure the model employs, during the training phase, about 370 s to complete a training epoch. These values were obtained by training the model on the Google Colaboratory platform. The trends of the training and the validation loss functions are shown in Figure 10.6.

After 60 training epochs, the model has reached a loss value of approximately 0.001 for both training and validation.

10.4.1　Data preparation

During data loading the 'generator' performs two fundamental steps for the functioning of the network. First of all, Sentinel-1 images, typically with a tif extension, are loaded into memory. Subsequently these undergo standardisation and normalisation so as to assume values comprised between 0 and 1; it is in fact known that the performance of a neural network strongly depends on the distribution of the values of the input data (recommended ranges are −1.1 or 0.1).

In fact, after the normalisation and standardisation process, speckle-type noise is added to the input image. Repeating this process for all the images of the dataset will obtain the vectors x training, y training, x validation and y validation with which the model will be trained. An example of this process is shown in Figure 10.7.

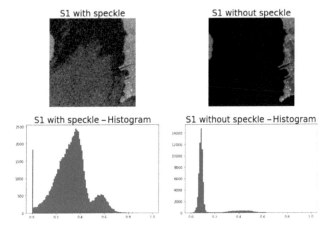

Figure 10.7   Example of the dataset

10.4.2   Test on small patches of a Sentinel-1 image

To evaluate the effectiveness of the filtering network, various tests have been carried out, both on the training set and on the testing set. The code has been made available to interested readers for further analysis and investigation [15].

10.4.2.1   Test on the training dataset

Typically, tests performed on training datasets lead to better results because the model somehow 'knows' the data and is therefore able to return the best possible result. Figure 10.8 shows three examples sorted by rows. For each line or example the filtered noise (residual noise), the input image with speckle noise, the image filtered by the network and the original image without speckle noise are reported.

As you can see from Figure 10.8 the model is able not only to extract a large amount of noise but also to preserve what is the structure of the image. To confirm these results in Table 10.2 the PSNR and SSIM values are reported.

In particular, the values of PSNR between the original image without the speckle and the image with the speckle were calculated and subsequently between the speckle-free image and the image filtered by the network. The same was done for the SSIM.

As you can see from the data reported in Table 10.2, the network is able to increase the SNR by about 14 and is able to perform good image restoration by raising the similarity index by about 0.6.

10.4.2.2   Test on validation dataset

The tests performed on the validation dataset show slightly lower results than in the previous case but this is part of the typical behaviour of any neural network model. Figure 10.9 shows three tests performed on the validation dataset.

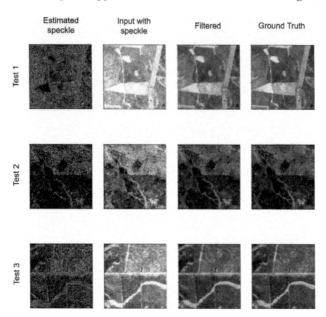

Figure 10.8   Test on training dataset

In general the model turns out to be really performing. To confirm these results in Table 10.3 the values of PSNR and SSIM are reported.

10.4.2.3   Comparison with the state of the art

The proposed model was found to be performing very well, in fact compared with the literature the network developed for the thesis work led to better results [12]. In Table 10.4, a comparison with the state of the art is presented.

Table 10.2   Evaluation metrics on training datasets

	Metrics	**Value**
Row 1	PSNR (clean, noise)	17.832 dB
	PSNR (clean, filtered)	31.805 dB
	SSIM (clean, noise)	0.374
	SSIM (clean, filtered)	0.910
Row 2	PSNR (clean, noise)	17.027 dB
	PSNR (clean, filtered)	31.946 dB
	SSIM (clean, noise)	0.303
	SSIM (clean, filtered)	0.901
Row 3	PSNR (clean, noise)	17.462 dB
	PSNR (clean, filtered)	31.527 dB
	SSIM (clean, noise)	0.361
	SSIM (clean, filtered)	0.897

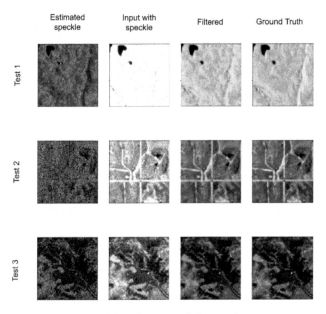

Figure 10.9   Test on validation dataset

10.4.3   Test on a whole Sentinel-1 image

As a final analysis, a script was created that was able to apply the filter created with the neural network on an entire Sentinel-1 image. In fact, so far, small patches of a Sentinel-1 image with a size of 256×256 pixels have been used. The algorithm created runs along the entire image with a window of 256×256 pixels. For each window the image is extracted and passed through the network to be filtered. During the application of the filter the input image is first normalised and standardised,

Table 10.3   Evaluation metrics on validation datasets

	Metrics	**Value**
Row 1	PSNR (clean, noise)	19.965 *dB*
	PSNR (clean, filtered)	32.357 *dB*
	SSIM (clean, noise)	0.654
	SSIM (clean, filtered)	0.926
Row 2	PSNR (clean, noise)	10.285 *dB*
	PSNR (clean, filtered)	25.796 *dB*
	SSIM (clean, noise)	0.379
	SSIM (clean, filtered)	0.791
Row 3	PSNR (clean, noise)	12.637 *dB*
	PSNR (clean, filtered)	28.344 *dB*
	SSIM (clean, noise)	0.364
	SSIM (clean, filtered)	0.813

Table 10.4 Evaluation metrics on validation datasets

Model	PSNR	SSIM
Input	14.53 *dB*	0.369
Lee	21.48 *dB*	0.511
Kuan	21.95 *dB*	0.592
PPB	21.74 *dB*	0.619
SAR-BM3D	22.99 *dB*	0.692
CNN	21.04 *dB*	0.631
SAR-CNN	23.59 *dB*	0.641
ID-CNN	24.74 *dB*	0.727
Proposed	28.83 *dB*	0.843

Batch with speckle   Batch filtered

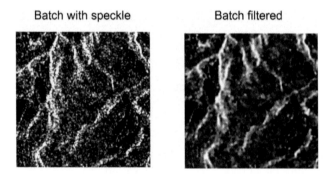

Figure 10.10 256×256 batch of a Sentinel-1 product

then filtered and then converted to bring it into initial conditions by applying the inverse standardisation and normalisation formulas and using the same weights (see Figure 10.10).

In Figure 10.11 you can see the original Sentinel-1 image with non-artificial speckle noise and in Figure 10.12 the filtered one.

The process of filtering the product took about 5 min to complete. Considering that the starting image is roughly formed by 25 000×16 000 pixels (approximately 6000 batches) the processing time is extremely low, especially when compared with that employed by a software like SNAP.

10.5 Conclusions

In this chapter, a possible CNN architecture applicable for speckle noise filtering has been presented. The results obtained are very promising, especially when compared with classical methods present in the state of the art.

The architecture presented can obviously be improved through a tuning of the hyperparameters. Among the future developments there is obviously the integration

Sentinel-1 product

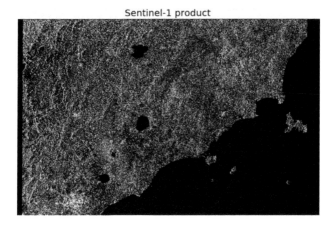

Figure 10.11   Original Sentinel-1 product with real speckle

Sentinel-1 product filtered

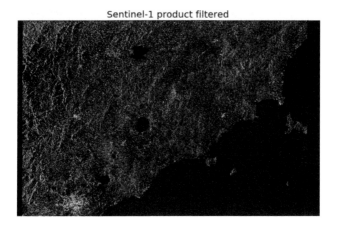

Figure 10.12   Filtered Sentinel-1 product without speckle

of a much more complex speckle model suitable for the content of the geographical portion contained in each single image of the dataset.

References

[1]   April G.V., Harvey E.R. 'Speckle statistics in four-look synthetic aperture radar imagery'. *Optical Engineering*. 1991;**30**(4):375–82.

[2]   Lee J.-S. 'Speckle suppression and analysis for synthetic aperture radar images'. *Optical Engineering*. 1986;**25**(5):255636.

[3]   Lopez-Martinez C., Fabregas X. 'Polarimetric SAR speckle noise model'. *IEEE Transactions on Geoscience and Remote Sensing*. 2003;**41**(10):2232–42.

[4]   Gagnon L., Jouan A. 'Speckle filtering of SAR images: a comparative study between complex-wavelet-based and standard filters'. *Wavelet Applications in Signal and Image Processing V*. **3169**. International Society for Optics and Photonics; 1997. pp. 80–91.

[5]   Quegan S., Le Toan T., Yu J.J., Ribbes F., Floury N. 'Multitemporal ERS SAR analysis applied to forest mapping'. *IEEE Transactions on Geoscience and Remote Sensing*. 2000;**38**(2):741–53.

[6]   Xiao J., Li J., Moody A. 'A detail-preserving and flexible adaptive filter for speckle suppression in SAR imagery'. *International Journal of Remote Sensing*. 2003;**24**(12):2451–65.

[7]   Peli T., Malah D. 'A study of edge detection algorithms'. *Computer Graphics and Image Processing*. 1982;**20**(1):1–21.

[8]   Irons J.R., Petersen G.W. 'Texture transforms of remote sensing data'. *Remote Sensing of Environment*. 1981;**11**:359–70.

[9]   Sebastianelli A., Del Rosso M.P., Ullo S.L. 'A SAR speckle filter based on residual Convolutional neural networks'. *arXiv preprint arXiv:2104.09350. Submitted to IEEE Geoscience and Remote Sensing Letters*. 2021.

[10] Sebastianelli A. 'SAR and optical data fusion using artificial intelligence. generation of Sentinel-2 images from despeckled Sentinel-1 data'. Thesis discussion date; 17-10-2019.

[11] Sebastianelli A., Del Rosso M.P., Ullo S.L. 'A SAR speckle filter based on residual ConvolutionalNeural networks'. *arXiv preprint arXiv:2104.09350*. 2021.

[12] Wang P., Zhang H., Patel V.M. 'Sar image despeckling using a convolutional neural network'. *IEEE Signal Processing Letters*. 2017;**24**(12):1763–7.

[13] Chierchia G., Cozzolino D., Poggi G. 'SAR image despeckling through convolutional neural networks'. International Geoscience and Remote Sensing Symposium (IGARSS); 2017.

[14] Sebastianelli A., Rosso M.P.D., Ullo S.L. Git-Hub Repository for SAR GRD Dataset. Available from https://github.com/Sebbyraft/sentinel_1_GRD_dataset [Accessed 6 July 2021].

[15] Sebastianelli A., Rosso M.P.D., Ullo S.L. Git-Hub Repository for SAR Speckle Filtering. Available from https://github.com/Sebbyraft/CNNSpeckleFilter [Accessed 6 July 2021].

Chapter 11

Future perspectives and conclusions

Silvia Liberata Ullo[1]

In this wrap up chapter we present possible perspectives of using AI (Artificial Intelligence) on board future satellite missions. Conclusions are briefly given at the end.

Some of our experiments will be presented and a first prototype introduced in this chapter. Moreover, considerations on the multidisciplinary aspects of the topics will be highlighted and other research fields listed, where the proposed AI and ML (Machine Learning) algorithms could be exploited.

11.1 Introduction

After the long journey through the several chapters of this book, the final question highlighted in this last chapter refers to what kind of future is forecast for AI and ML algorithms in the field of satellite remote sensing.

As already highlighted, after the creation of a good detection model for the observation of a phenomenon of interest, its optimisation in terms of network complexity, number of parameters and execution time to make a new prediction, becomes the next issue to focus on. These specifics have acquired a great importance for the research: the idea is to develop and apply a classification model on a small chip, bonded to the elaboration power of the chip itself, and the on-board computer.

The case study proposed in this chapter refers to the simulation of a satellite with AI on board. For this purpose, the algorithms developed for the detection of volcanoes, presented and discussed in Chapter 8, will be utilised. In particular, the on-board computer will be simulated by a Raspberry Pi board to which the data acquisition system, an RGB (Red-Green-Blue) camera, and the chip for the neural network, the Movidius Stick, are connected.

[1]Professor at Engineering Department, University of Sannio, Benevento, Italy

11.2   A first prototype

The prototype, as mentioned earlier, is implemented on a Raspberry Pi, using the Movidius Stick to run the neural network. The case study presented in this section is an evolution of the volcanic eruption binary classification problem presented in Chapter 8.

11.2.1   Raspberry Pi

The Raspberry Pi is a series of small single-board computers developed in the United Kingdom by the Raspberry Pi Foundation to promote teaching of basic computer science in schools and in developing countries. The original model became far more popular than anticipated, selling outside its target market for uses such as robotics. It does not include peripherals (such as keyboards and mice) or cases. However, some accessories have been included in several official and unofficial bundles.

The Raspberry Pi Foundation provides **Raspbian**, a Debian-based Linux distribution for download. It promotes Python and Scratch as the main programming languages, with support for many other languages. The default firmware is closed source, while an unofficial open source is available. Many other operating systems can also run on the Raspberry Pi, including the formally verified microkernel seL4. Other third-party operating systems available via the official website include Ubuntu MATE, Windows 10 IoT Core, RISC OS and specialised distributions for the Kodi media centre and classroom management. To install the operative system a Micro SD card is needed, with New Out Of the Box Software (**NOOBS**), an easy operating system installer which contains Raspbian and LibreELEC. It also provides a selection of alternative operating systems which are then downloaded from the internet and installed. The model used for the case study is the **Raspberry Pi 3 B**, shown in Figure 11.1.

Figure 11.1   Raspberry Pi 3 B (Source[1])

Table 11.1   Raspberry PI 3 specifications

Quad Core 1.2 GHz Broadcom BCM2837 64 bit CPU 1GB RAM (Random Access
  Memory)
BCM43438 wireless LAN (Local Area Network) and Bluetooth Low Energy (BLE) on
  board
100 Base Ethernet
40-pin extended GPIO (General Purposed Input-Output)
4 USB (Universal Serial Bus) 2.0 ports
4 Pole stereo output and composite video port
Full size HDMI (High-Definition Multimedia Interface)
CSI (Camera Serial Interface) camera port for connecting a Raspberry Pi camera
DSI (Display Serial Interface) display port for connecting a Raspberry Pi touchscreen
  display
Micro SD (Secure Digital) port for loading your operating system and storing data
Upgraded switched Micro USB power source up to 2.5 A

The Raspberry Pi 3 Model B is the earliest model of the third-generation
Raspberry Pi (Table 11.1 shows its main specifications).

11.2.2   Movidius Stick

Intel Movidius Neural Compute Stick (NCS) is a tiny fanless deep learning USB
drive (shown in Figure 11.2), designed to learn AI programming. The NCS is powered by the low power high performance Movidius vision processing unit (VPU).
The VPU can be found in millions of smart security cameras, gesture controlled
drones, industrial machine vision equipment, etc.

The main features of the chip can be summarised as:

* Supporting convolutional neural network (CNN) profiling, prototyping and tuning workflow

Figure 11.2   Movidius stick (Source [2])

- Real-time on device inference (cloud connectivity not required)
- Features the Movidius VPU with energy-efficient CNN processing
- All data and power provided over a single USB type A port
- Run multiple devices on the same platform to scale performance.

11.2.3 Camera

The Raspberry Pi Camera Module v2, in Figure 11.3, is a high quality 8 megapixel Sony IMX219 image sensor custom designed add-on board for Raspberry Pi, featuring a fixed focus lens. It is capable of 3280x2464 pixel static images, and supports 1080p30, 720p60 and 640x480p90 video.

It attaches to Pi by way of one of the small sockets on the board upper surface and uses the dedicated CSi interface, designed especially for interfacing to cameras. The main features are listed in Table 11.2.

11.2.4 Proposed model

For the construction of a smaller network, which preserves the qualities proposed in Chapter 8, various strategies have been adopted [4]. The first step was to try reducing the size of the network as much as possible. In fact, knowing that most of the parameters were depending on the number of filters and convolutional layers used, their drastic reduction was decided.

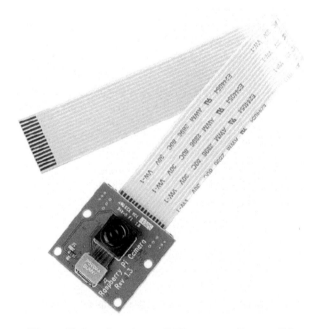

Figure 11.3   Raspberry RGB camera (Source [3])

Table 11.2   Raspberry Pi RGB camera features

Fixed focus lens on-board
8Megapixel native resolution sensor-capable of 3280×2464 pixel static images
Supports 1080p30, 720p60 and 640 × 480p90 video
Size 25 mm×23 mm × 9 mm
Weight just over 3 g
Connects to the Raspberry Pi board via a short ribbon cable (supplied)
Camera v2 is supported in the latest version of Raspbian

In this case the network will be able to abstract at a lower level than the previous version. If you think about the possible features that the net can extract (the crater's circular shape, lava courses, etc.), it is plausible to think that even a network with a lower number of layers could be sufficient. In fact, at the cost of greater instability, the new network is still able to discriminate or classify data correctly most of the times. This approach could not be feasible for a different case study. All changes are implemented in order to reduce the number of network parameters: the faster the model, the faster the acquisition and the processing of the images.

11.2.5   Results

The model has been trained for 100 epochs, using the **Adam optimiser** and the **binary cross-entropy** loss function. The trends of the training and the validation loss and accuracy functions are shown in Figure 11.4.

Compared with the model presented in Chapter 8, the accuracy values on training and validation datasets are lower, but with a slight fluctuation they maintain the same trends. On the other hand, the trends of validation and training loss are very different and the validation loss presents great oscillations especially in the second part. This highlights the fact that the model is not big enough to make the right predictions, also because the validation loss changes a lot between epochs The confusion matrix is shown in Figure 11.5.

The model with the highest value of validation accuracy has been chosen through the early stopping technique. For this reason the results of predictions on the validation dataset are quite good. Considering the different numbers of images in the two classes, the percentage of right and wrong classifications is the same for both, the eruption and the no eruption class. Of the images 83% are classified correctly and the other 17% are classified as the opposite class.

11.2.6   Implementation on Raspberry and Movidius Stick

The implementation process is shown in Figure 11.6.

First of all two preliminary steps should be followed.

- The network has to be converted from the Keras model to an OpenVINO format, using Intel OpenVINO library. This step is essential to make it run on the Movidius Stick.

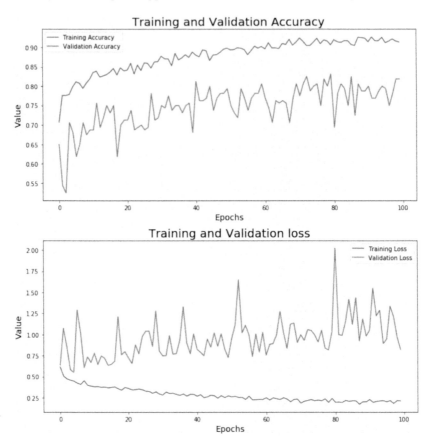

Figure 11.4   Accuracy and loss functions for the small model

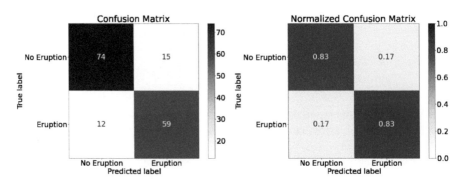

Figure 11.5   Confusion matrix for the small model

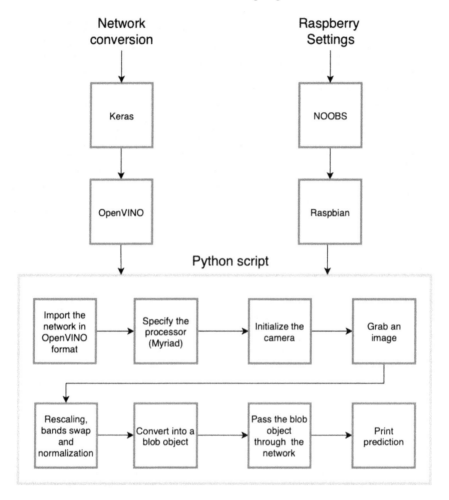

Figure 11.6   Block diagram for the implementation on chip

- An operative system should be installed on the Raspberry. In this case the Raspbian Operative System has been installed using the NOOBS installer.

11.2.7   OpenVINO library

For deep learning, the current Raspberry Pi hardware is inherently resource-constrained. Intel's Movidius NCS allows for faster inference with the deep learning co-processor that is plugged into the USB socket. In order to transfer the network on the Movidius, the network should be optimised. For this purpose, **OpenVINO**, an Intel library for hardware optimised computer vision, will be used.

OpenVINO is extremely simple to use: you just need to set the target processor (a single function call) and let OpenVINO-optimised OpenCV handle the rest.

To use OpenVINO you need:

- Raspberry Pi 3B+ (or Raspberry Pi 3B)
- Movidius NCS 2 (or Movidius NCS 1)
- PiCamera V2 (or USB webcam)
- 32 GB microSD card with Raspbian Stretch freshly flashed (16 GB would likely work as well)
- HDMI screen + keyboard/mouse (at least for the initial WiFi configuration)
- 5 V power supply (I recommend a 2.5 A supply because the Movidius NCS is a power hog).

The OpenVINO toolkit is an Intel R distribution. Based on CNNs, the toolkit extends workloads across Intel R hardware (including accelerators) and maximises performance:

- Enables deep learning inference at the edge
- Supports heterogeneous execution across computer vision accelerators (CPU (Central Processing Unit) Proc), GPU (Graphics Processing Units), Intel R Movidius NCS and FPGA (Field Programmable Gate Array)) using a common API (Application Program Interface)
- Speeds up time to market via a library of functions and preoptimised kernels
- Includes optimised calls for OpenCV and OpenVX.

11.3   Future satellite missions with AI on board

As highlighted at the beginning of this chapter, the next step in the Earth Observation science is the creation and launch of satellites with artificial intelligence on board. AI will provide great improvement in preprocessing and managing satellite data, by providing opportunities ever thought before.

A satellite with an AI computer on board would be able to decide how much data to send to the ground stations and how much to process locally. Specific AI algorithms could provide specific preprocessing steps as image filtering, data compression, geometric correction, radiometric correction, atmospheric correction and other image transformations.

A higher level of data processing could include the managing of the data for specific applications, for example, the monitoring of volcanic eruptions or other natural disasters. With this aim, the satellite could include:

- Onboard thematic classifier for disaster warning and monitoring
- A change detection system in order to transmit only specified changed data.

Figure 11.7   φ-Sat chip (Source [6])

11.3.1   ESA first intelligent satellites: φ -Sat 1 and 2

On the 3 September 2020, the first European intelligent satellite has been launched. Its purpose is to demonstrate how onboard AI can improve the efficiency of sending earth observation data back to earth. Dubbed φ-Sat, or PhiSat, this revolutionary AI technology has been designed to fly on one of the two CubeSats that make up the FSSCat mission, a Copernicus Masters winning idea [5]. As the overall 2017 Copernicus Masters winner, FSSCat was proposed by Spain Universitat Politècnica de Catalunya and developed by a consortium of European companies and institutes.

The two CubeSats, each about the size of a shoebox, collect data, available through the Copernicus Land and Marine Environment services, using state-of-the-art dual microwave and hyperspectral optical instruments. They also carry a set of intersatellite communication technology experiments.

The hyperspectral camera on one of the CubeSats will collect an enormous number of images of the earth, some of which will not be suitable for use because of cloud cover. To avoid downlinking these not perfect images back to earth, the φ-Sat artificial intelligence chip will filter them out so that only usable data will be returned (see Figure 11.7).

After the success of the Φ-Sat 1, ESA published a call for a second AI satellite and selected an idea [7] concerning 6U CubeSat platform capable of running AI apps that can be developed, easily installed, validated and operated on the spacecraft during their flight using a simple user interface.

The mission will address a wide range of AI applications including: transforming a satellite image to a street map, cloud detection in order to reduce the volume of data to be downlinked to the ground, autonomous detection and classification of maritime vessels, as well as forest monitoring and anomaly detection.

The spacecraft will fly in a sun-synchronous orbit with a powerful multispectral optical camera with an expected 4.75 m ground resolution. The payload will include the AI processor Intel Movidius Myriad 2 from Ubotica, which was already adopted on the Φ-Sat 1 mission.

It is worth highlighting that, in order to demonstrate the potential of AI in space, ESA has been working heavily during these last years with some partners to develop Φ-Sat and take the mission to the next level .

11.4   Conclusions

Looking back at all the chapters and topics discussed we really hope to have created a useful guide for many researchers interested in understanding how machine learning algorithms can be applied to solve many problems related to satellite remote sensing. Our contacts are in the book and we look forward to receiving any feedback and request for clarification.

11.5   Acknowledgements

In conclusion of this book, we want to thank Giancarlo Filippazzo of ESA (European Space Agency) ESRIN (European Space Research Institute), who made possible our first visit to ESA headquarters in Frascati, in January 2019. Thanks to him, the contact between the University of Sannio and the ESA Φ-Lab was born.

A special thanks to Pierre Philippe Mathieu, head of the ESA Φ-Lab Explore Office, whom we have esteemed since the very first moment, for the passion he puts in his job.

A special thanks also to Dario Spiller, Andrea Radius, Erika Puglisi and Sara Aparicio for actively working on the book.

We also want to thank all the coauthors who, working on the various chapters, contributed to the realisation of this book: Artur Nowakowski, Chris Stewart, James Wheeler, Chiara Zarro, Filippo Biondi, Pia Addabbo, Carmine Clemente, Danilo Orlando, Diego Di Martire, Mariano Di Napoli, Mario Luca Bernardi, Marta Cimitile, Rogerio Bonifacio, Do-Hyung Kim and Manuel Garcia-Herranz. We also want to thank Raquel Carmo, Noelle Cremer, Dario Amorosi and Dawa Derksen, who participated in the revision of some chapters.

Lastly, but not the least, our big thanks go to ESA, to its vision of openness not only towards the free accessibility of satellite data and tools, but also towards the promotion of several opportunities of collaboration and knowledge exchange. The ESA Φ-lab, in particular, is an awesome dream site where cross-fertilization of ideas and cutting-edge topics takes place with people coming from different countries worldwide.

References

[1] *Raspberry Pi 3 B board [online]*. Available from www.raspberrypi.org/prod-ucts/raspberry-pi-3-model-b/ [Accessed October 2020].

[2] Stick M. Intel Movidious Neural Compute Stick. Available from https://ark.intel.com/content/www/it/it/ark/products/125743/intel-movidius-neural-com-pute-stick.html [Accessed October 2020].

[3] *Raspberry RGB camera [online]*. Available from www.raspberrypi.org/prod-ucts/camera-module-v2/ [Accessed October 2020].

[4] Rosso M.P.D. 'Convolutional neural networks for the detection of volcanic eruptions: towards a new generation of satellites with artificial intelligence on board'. Thesis discussion date; 17-10-2019.

[5] ESA. *FSSCat CubeSats [online]*. Available from www.esa.int/ESA Multimedia/ Images/2020/05/FSSCatCubeSats [Accessed October 2020].

[6] *φ-Sat 1 chip [online]*. Available from https://visionspace.blog/blog/2020/8/7/ phi-sat-1-and-its-ai-chip [Accessed October 2020].

[7] ESA. *φ-Sat 2 idea*. Available from www.esa.int/Applications/_Observing_the_ Earth/Ph-sat/Next_artificial_intelligence_mission_selected [Accessed October 2020].

Index

absorption 18
accelerometer 20
active remote sensors
  example 19
  laser altimeter 19–20
  LiDAR 18–19
  RADAR 20
  ranging instrument 20
  scatterometer 20
  sounder 20
adaptive filters 223
adaptive pooling, convolutional layers 107
Adaptive Synthetic Sampling Method (ADASYN) 53
airborne platform- based remote sensing 22–3
AlexNet-based network 91
artificial intelligence (AI)
  adaptability 40
  Dartmouth Summer Research Project 41
  data-intensive tasks 45
  deep learning 43–4
  deep neural network 42
  history 40
  intelligence 40
  intentionality 40
  machine learning 43
  real-life applications 44–5
  remote sensing applications 55–7
  repetitive tasks 45
  revolution 2
  simple neural network 42
  superhuman tasks 45

Artificial Intelligence for Earth Observation (AI4EO)
  adaptative data engine 3–4
  AI@edge 4–6
  AI inference 6
  AI processing 3
  building capacity 6
  deep learning (DL) 3
  Digital Twin Earth 6, 7
  frontier scientific problem 6
  machines learning algorithms 3
artificial neural network (ANN)
  activation functions
    linear function 66
    non-linear function 66–72
    step function 66
  artificial neurons 64, 65
  forward and backward propagation (see forward and backward propagation)
  generalisation 63–4
  neural network 64
  overfitting and underfitting 84–7
  propagation function 65
  regularisation techniques 87–8
artificial neurons 64, 65
atmospheric correction, data quality 26
automatic differentiation technique 80
average pooling, convolutional layers 107
azimuth spatial resolution 28

back propagation
  automatic differentiation 80, 83

differentiation 82–3
forward propagate 81
iterations 84
learning process 79
loss functions 81–2
model initialisation 81
prediction process 79
weight update 83–4
bagging 86
batch normalisation 77–8, 107–8
bias and variance 84–5
BigEarthNet project 195
binary cross-entropy formula 164
bounding box 127

chessboard segmentation 196
classification problem
    crop-type classification (CTC)
        191–5
    landslides binary classification (*see*
        landslides binary classification)
    segmentation 195–9
    volcanic eruptions (*see* volcanic
        eruptions binary classification)
class imbalance 52–3
cloud masking 32
contrast split segmentation 196–7
convolutional neural network (CNN/
    ConvNet)
    batch normalisation 107–8
    classification task 108
    convolution operation
        commutative operator 96
        cross-correlation 97
        discrete convolution 95
        functional analysis 95, 96
        multidimensional arrays 95–6
        object position 95
        two-dimensional convolution
            formula 96
    crop-type classification (CTC)
        193–4
    detection 109
    feature extraction phase 92

fully-connectedness 91
hidden layers 93
image segmentation task 109
kernels and feature maps
    convolution operation 100
    edge detectors 101
    multidimensional input 103
    padding 103–4
    stride 102–3
    VGG16 network 101, 102
landslides binary classification
    172–3
motivation 93–4
multilayer perceptrons 91
networks and layers structure 99–100
object localisation task 109
parameter sharing
    edge detection 98
    equivariance 98
    function mapping 98
    time series data processing 98–9
    traditional neural network 97
rectified linear unit layer 104–6
Sentinel-1 speckle filtering 226–7
space invariant artificial neural
    network 92
sparse connectivity 93–4
transfer learning 109
Copernicus programme 9
    optical sensors 31–4
    Sentinels 26–7
    synthetic aperture radar (SAR)
        28–33
crop-type classification (CTC)
    BigEarthNet project 195
    CNNs and transfer learning 193–4
    crop mask 191
    dataset 192–3
cross-entropy loss 164
CubeSats 245

data augmentation 53–4
    binary classification problem 53
    classifier 53

dataset choice and use 54–5
  image data 54
  tabular data 53
data-intensive tasks 45
debris flow and slides 174
deep generative models
  generative adversarial networks
    210–11
  variational autoencoders (VAEs)
    209–10
deep learning (DL) 2, 43–4
deep learning model implementation
  importing libraries 138, 144
  noise filtering
    generator 145
    model building 145–7
    model train and test 147–8
  object detection
    generator 153–4
    model building 154–5
    model train and test 155–7
  objects classification
    generator 148–50
    model building 150–51
    model train and test 151–3
deep neural network 42, 64
digital elevation model (DEM) 31
digital revolution 1
digital surface models (DSMs) 197
digital terrain models (DTMs) 197
Digital Twin Earth (DTE) framework
  6, 7
Doppler effect 14–15
dropout 86–7

earth flow 174
eCognition, segmentation algorithms
  196
edge based segmentation 196
electromagnetic radiation
  Doppler effect 14–15
  electromagnetic wave 10–11
  fluorescence 16
  interaction with atmosphere

    absorption 18
    refraction 18
    scattering 17
  polarisation 16
  spectrum 11–12
  transmission 15–16
  wave and matter interaction
    diffraction 14
    reflection 12–13
    refraction 13–14
equivalent number of looks (ENL) 225
European intelligent satellite 245

feedforward neural network 64
filtering problem. *See* speckle filtering
fit generator 147
fluorescence 16
forward and backward propagation
  back propagation 79–84
  batch normalisation 77–8
  critical or stationary points 73
  directional derivative 74–5
  full batch gradient descent 76
  gradient descent 72–3
  gradient explosion 79
  line search 75
  mean absolute error 72
  mini-batch gradient descent 76–7
  minimisation 74
  objective function 73
  optimisation algorithms 74
  stochastic gradient descent (SGD)
    75–6
  vanishing gradient 78
FSSCat mission 245
full batch gradient descent 76

generation problem
  deep generative models 208–11
  generative networks 207
  optical-like images generation
    dataset and proposed methods
    212

discriminator 213, 214
generator 213, 214
loss function selection 213–14
test model 214–18
generative adversarial networks
(GANs) 210–11
geocoding 30–1
global pooling, convolutional layers
107
Google Colaboratory
Colab Pro 135
Google Drive connectivity 136–8
graphical user interface (GUI) 136
graphics processing unit 135–7
Google Earth Engine (GEE)
online editor 117–18, 125–7
Python API 119–20
GoogLeNet 193
gradient descent 72–3
graphical user interface (GUI) 136
ground-based platforms 21–2
ground-based synthetic aperture radar
interferometry (GB-InSAR)
172
ground range detected (GRD) images
29

hyperbolic tangent function 67–8
hyperspectral radiometer 21

image loader 165
infrared radiation 11
interferometric synthetic aperture radar
(InSAR) 32
landslides binary classification
171–2

j-Sat 1 and 2 245

Keras 132, 144
k-fold cross-validation 50, 51

k-means clustering 48

LabelImg image annotation tool 128–9
landslides binary classification
automated detection 171
confusion matrix network 187, 188
convolutional neural networks
optical images 176–80
Sentinel-1 images 173
Sentinel-2 images 173, 176–80
strengths 172
dataset maintenance 190
geological characteristics 174–6
InSAR techniques 171–2
Keras accuracy 186
landslide classification 173–4
main landslides types 174, 175
neural network classifier 189
RGB composite summary scheme
190–1
SAR data 180–81
change map algorithm 181
Colab platform 185
Ground Range Detected (GRD)
product 181
image generator 185–6
landslides 182, 183
proposed CNN architecture 184–5
Sentinel-1 SAR image 182, 183
Single Look Complex (SLC) 181
training dataset 186
validation dataset 186
sliding and falling masses 171
training and validation accuracy 187
laser altimeter 19–20
Lee filter 224
light detection and ranging sensor
(LiDAR) 18–19
linear activation function 66
Linux/MacOS 142–4

machine learning (ML)
algorithms 3, 43

class imbalance 52–3
data augmentation 53–5
e-commerce sites and social media
    57
edge device 4–5
fault detection, navigation systems
    57–8
medical applications 58
model training and validation 50, 51
self-driving cars 57
space exploration 57
supervised learning 45–7
unsupervised learning 47–8
makesense.ai 121–4
Matplotlib 144
max pooling, convolutional layers
    106–7
Meteosat Third Generation- Sounder
    (MTG-S) satellite 27
microwave radiation 11
Mie scattering 17
min pooling, convolutional layers 107
Movidius Stick 239–40
multilook detected (MLD) images 29
multilook processing 223
multi-modal speckle filtering 225–6
multiresolution segmentation (MRS)
    196–9
multispectral sensor 20
multitemporal filtering 224

New Out Of the Box Software
    (NOOBS) 238
non-adaptive filters 223
non-linear activation functions
    bias 69–72
    hyperbolic tangent function 67–8
    rectified linear unit function 68–9
    sigmoid function 67
nonselective scattering 17
normalised difference vegetation index
    (NDVI) 27
normalized digital surface models
    (NDSMs) 197

NumPy 138

object-based image analysis (OBIA)
    33–4, 195
Ocean and Land Colour Instrument
    (OLCI) 27
off-policy method 50
one-input one-output neural network
    71
on-policy method 50
open access hub 114–16
OpenVINO library 243–4
optical sensors
    atmospheric corrections 32
    classic processing workflows 31–2
    cloud masking 32
    indices 32
    object-based image analysis
        technique 33–4
    radiance 31
    re-sampling 31
orbit file refining 29–30
overfitting and underfitting
    bias and variance 84–5
    dropout 86–7
    early stopping of training 85–6

passive remote sensors 19–21
peak signal-to-noise ratio (PSNR) 228
persistent scatterer interferometry
    (PSI) 172
point based segmentation 196
polarimetry 223
polarisation 16
pooling, convolutional layers 106–7
principal component analysis (PCA)
    47–8
propagation function 65
pruning, neural network 5
Python 119–20, 143–8

Q-learning 49

quadtree segmentation 196

radiance 31
radio detection and ranging sensor
(RADAR) 20
radiometer 21
radiometric correction, data quality 25
radiometric resolution 24, 25
Raspberry Pi
accuracy and loss functions 241,
242
Adam optimiser 241
camera 240, 241
confusion matrix, small model 242
implementation process 241, 243
Movidius Stick 239–40
OpenVINO library 243–4
operating systems 238
proposed model 240–41
Raspberry Pi 3 B 238
specifications 239
Rayleigh scattering 17
rectified linear unit (ReLU) 68–9
activation functions 105
interactions 105–6
mathematical formulation 104
non-linearities 106
recurrent neural network 64
reflection 12–13
refraction 13–14, 18
region based segmentation 196
regularisation techniques 87–8
reinforcement learning (RL) 49–50
remote sensing
data characteristics
data quality improvement 25–6
radiometric resolution 24, 25
spatial resolution 23–4
spectral resolution 24
temporal resolution 24–5
electromagnetic radiation
Doppler effect 14–15
fluorescence 16

interaction with atmosphere 16–18
polarisation 16
principles 10–12
transmission 15–16
waves and matter interaction 12–14
elements 9
frequency intervals 11
natural disasters 55–7
platforms 21–3
sensors
active 18–20
passive 19–21
rock slide and falls 174

satellite data analysis
Copernicus programme 26–7
remote sensing (*see* remote sensing)
scattering 17
scatterometer 20
segmentation
categories 195, 196
contrast split segmentation 196–7
deep learning and object-based
analysis integration 199
edge based 195
multiresolution segmentation 197–9
object-based image analysis (OBIA)
195
pixel-based classification 195
point based 195
region based 195
tool and algorithms 196
semantic segmentation tasks
Google Earth Engine online editor
125–7
LabelImg 128–9
training dataset 123–5
sensing revolution 1
sensors
active 18–20
passive 19–21
Sentinel Application Platform (SNAP)
116–17

Sentinels 26–7
sigmoid function 67
simple neural network 42
single look complex (SLC) data 28–9
slide spreads 177
Snell's law 14
spaceborne platforms 23
space invariant artificial neural network 92
sparse connectivity, deep convolutional network 93–4
spatial resolution 23–4
speckle filtering
  adaptive, 223
  evaluation metrics
    peak signal-to-noise ratio (PSNR) 228
    structural similarity index 229
    validation datasets, 234
  motivation 222–3
  non-adaptive 223
  reduction techniques 223–4
  Sentinel-1
    CNN model 226–7
    data preparation 230
    filtered product without speckle, 235
    loss function selection 228
    multi-modal speckle filtering 225–6
    network architecture 227
    pure SAR strategy 226, 227
    training dataset 231–2
    validation dataset 232, 233
  speckle filters 224
  speckle noise 221
  synthetic aperture radar (SAR) image 222
spectral difference segmentation 200
spectral resolution 24
spectrometer 21
spectroradiometer 21
state–action–reward–state–action (SARSA) 50

step activation function 66
stochastic gradient descent (SGD) 75–6, 108
structural similarity index (SSIM) 229
superhuman tasks 45
supervised learning
  algorithms 46
  cross-validation 50–2
  objective 46
  problem solving steps 46–7
synthetic aperture radar (SAR)
  azimuth spatial resolution 28
  classic processing workflows
    geocoding 30–1
    levels 29
    multilooking processing 30
    orbit file refining 29–30
    radiometric calibration 30
    radiometric terrain flattening 30
    specklefiltering 30
    thermal noise removal 30
  interferometry 32–3
  multilook detected (MLD) images 29
  principles 28
  single look complex (SLC) data 28–9
Synthetic Minority Oversampling Technique (SMOTE) 53

Tanh function 67–8
temporal resolution 24–5
TensorFlow
  CPU-only version 140–41
  features 131
  graphics processing unit
    AMD GPUs 144
    NVIDIA GPUs 141–4
  Keras 132
  libraries 132
  Linux/MacOS 142–4
  Nvidia CUDA Toolkit 141–2
  programming paradigm 131

virtual environment creation 139–40
tensors 96
    dtype 132
    multiple dimensions 133
    rank 133
    three-axess tensor, shape 134
topographic correction, data quality
        25–6
total internal reflection law 13
training datasets
    automated method 117–20
    dataset annotation 120–21
    image classification tasks, 121–3
    non-automated method
        open access hub 114–16
        Sentinel Application Platform
            116–17
    object detection tasks 127–9
    quality 114
    quantity 114
    representative 113–14
    semantic segmentation tasks 123–7
transfer learning 109, 193–4
transmission of radiation 15–16
TROPOspheric Monitoring Instrument
        (TROPOMI) 27
two-dimensional convolution formula
        96

unsupervised learning 47–8

valid padding strategy 104
vanishing gradient 78
variational autoencoders (VAEs)
        208–10
VGG16 193
visible radiation 11
volcanic eruptions binary classification
    catalog 159–60
    data preparation and manipulation
            160–61
    dataset expansion 161–2
    proposed model
        convolutional sub-network 162–3
        cross-entropy loss 164
        dense sub-network 164
        image loader 165
        network architecture 164
    Sentinel-5P data 171, 172
    test dataset 167–70
    training dataset
        accuracy and loss functions 165,
            166
        confusion matrix 166, 167
        graphics processing units (GPUs)
            165, 166